# The Temporalization of Time

# The Temporalization of Time

*Basic Tendencies in Modern Debate on Time in Philosophy and Science*

Mike Sandbothe

Translated by Andrew Inkpin

ROWMAN & LITTLEFIELD PUBLISHERS, INC.
Lanham • Boulder • New York • Oxford

ROWMAN & LITTLEFIELD PUBLISHERS, INC.

Published in the United States of America
by Rowman & Littlefield Publishers, Inc.
4720 Boston Way, Lanham, Maryland 20706
www.rowmanlittlefield.com

12 Hid's Copse Road
Cumnor Hill, Oxford OX2 9JJ, England



British Library Cataloguing in Publication Information Available

**Library of Congress Cataloging-in-Publication Data**

Sandbothe, Mike, 1961–
[Verzeitlichung der Zeit. English]
The temporalization of time : basic tendencies in modern debate on time in philosophy and science / Mike Sandbothe.
p. cm.
Includes bibliographical references and index.
ISBN 0-7425-1289-4 (alk. paper) — ISBN 0-7425-1290-8 (pbk.: alk. paper)
1. Time. I. Title.
BD638.S27
115—dc21 2001041696

∞™ The paper used in this publication meets the minimum requirements of American National Standard for Information Sciences—Permanence of Paper for Printed Library Materials, ANSI/NISO Z39.48–1992.

# Contents

# Acknowledgments

This work is a revised and abridged version of my doctoral thesis. I would like to thank both supervisors of the work, Wolfgang Welsch and Walther Ch. Zimmerli, for their constructive criticism. The same applies to Ilya Prigogine and Isabelle Stengers. In our meetings they not only helped me out on detailed matters in physics and the history of science but also were open to controversial discussions on philosophical questions. Andrew Inkpin translated the work from German to English. He also gave me some important suggestions with regard to content. I would like to thank him for both. Without Joe Margolis this book probably would not have appeared in English. I would like to thank him for his support. It has been a pleasure to work with Eve DeVaro, Whitney Poole, and Julie Kirsch of Rowman & Littlefield. The book is dedicated to my parents Theodor (†) and Hildegard Sandbothe.

# Introduction

In the current situation, one characterized by a plurality of heterogeneous time concepts, cross-disciplinary discussion about the problem of time assumes particular importance.[1] The central problem for current debate on time is to relate the varying concepts of time developed in individual scientific disciplines both to one another and to everyday experience (cf. Baert, 1999; Baumgartner, 1993; Burger, 1993; Gimmler, Sandbothe, and Zimmerli, 1997; Le Poidevin and McBeath, 1993; Mainzer, 1996).[2] In the attempt to solve this task different approaches can be distinguished. They are embedded in two lines of development, to be outlined below, which determine current time theory.

The first basic tendency in contemporary philosophy of time may be described as the tendency *to unify and to universalize* our understanding of time. The protagonists of this tendency are convinced that the aspect of time is to be considered a new Archimedean point, unifying our everyday experience of self and the world with scientific theories about humankind and nature. This point of unity, they contend further, has been highlighted over and over again in philosophy (for instance by von Baader, Schelling, Bergson, Whitehead or Heidegger), but has been ignored for far too long by science and technology. It was not until the second half of the twentieth century that a global time concept was developed and mathematically operationalized at the interface between physics, chemistry and biology within the framework of the so-called theories of "self-organization" (cf. Griffin, 1986; Krohn, Küppers, and Nowotny, 1990). According to the proponents of the unification tendency, this new conception of time enables the old duality between natural time and historical time to be overcome and marks the beginning of the resolution of the conflict between physical and philosophical thinking about time which had been characteristic of time theories at the start of the twentieth century. In this sense the German philosopher of time and history Hermann Lübbe observed in his book *In the Course of Time* "that even the temporal structure of historicality, which, according to Heidegger and the hermeneutic theory that followed him, results exclusively from the subject's relationship to itself and its constituting of meaning, is in reality a structure indifferent to subject matter, belonging to all open and dynamic systems" (Lübbe, 1992, p. 30).

Lübbe's convergence theorem can draw support from the deliberations of one of the founders of self-organization theory. Already in 1973, the Nobel prize–winning physicist and chemist Ilya Prigogine noted with his theory of irreversible structures in mind: "Whatever the future of these ideas, it seems to me that the dialogue between physics and natural philosophy can begin on a new basis. I don't think that I can exaggerate by stating that the problem of time marks specifically the divorce between physics on one side, psychology and epistemology on the other. . . . We see that physics is starting to overcome these barriers" (Prigogine, 1973, p. 590f.). Prigogine further developed the specific signature of the current debate on time in the closing chapter of his 1984 revision to the German edition of *Being and Becoming*:[3] "It is remarkable to recognize the extent to which some of the recent results [of natural science—M.S.] had been anticipated by philosophers like Bergson, Whitehead and Heidegger. The main difference consists of the fact that they could reach such conclusions only in contrast to natural science, whereas we are now observing that these insights emerge so to speak from scientific research" (Prigogine, 1988, p. 262). Prigogine's convergence theorem is found again, more precisely formulated, in an essay he published together with Serge Pahaut in 1988:

> Both classical and relativistic or quantum physics concentrated on time considered as motion. It seemed as if time, as qualitative change, lie outside its horizon. From this there results on one side the temptation, which we meet even with Einstein, to deny the existence of time or history, and on the other side there result from this the objections of philosophers like Bergson, Whitehead, Husserl or Heidegger, who see the pauper's oath of the scientific method in this denial. Strangely enough we can today set our sights on the possibility of a synthesis linking these two aspects of time with each other. (Prigogine and Pahaut, 1985, p. 26)

The second line of development in contemporary theory of time is best seen when one reconsiders the assumptions common to the advocates of the unification and universalization tendency. Time is considered by them to be a uniform universal base structure which disavows itself of historical contingency and cultural change. Thus Lübbe and Prigogine consider the "ontological universality of the temporality aspect" (Lübbe, 1992, p. 31) of self-organization's "participatory universe" (Prigogine and Stengers, 1981, pp. 267ff., 287f.; cf. also Wheeler, 1979, pp. 407ff.) to be evident. Advocates of the second basic tendency, a tendency to *historize and relativize* time, proceed from the basic idea that the role played by time in human understanding of self and the world is one aspect of a system of practical and technical habits which diverges between cultures and changes within a culture in contingent conditions over history.

This approach is advocated with particular refinement by the American pragmatist Richard Rorty. The basic premise of Rorty's thinking is "that a belief can still regulate action, can still be thought worth dying for, among people who are quite aware that this belief is caused by nothing deeper than contingent historical circumstances" (Rorty, 1989, p. 189). According to Rorty radically temporal thinking must do away with the theologically founded conception that time and eternity come together in man (Rorty, 1995). Instead Rorty demands "that we [should] try to get to the point where we no longer worship *anything*, where we treat *nothing* as a quasi divinity, where we treat *everything*—our language, our conscience, our community—as a product of time and chance" (Rorty, 1989, p. 22). According to Rorty we will succeed in this only when we no longer mystify time, but understand it in a radically reflexive way as being a product of chance (Gimmler, Sandbothe, and Zimmerli, 1997, pp. 1–78; cf. Janich, 1996).

The interrelations between the different concepts of time currently being discussed in the sciences, as well as the question of the relationship between academic and everyday perceptions of time, are to be dealt with pragmatically on the basis of the historization tendency advocated by Rorty. Convergence between different vocabularies of time is, from Rorty's perspective, by no means proof of an intrinsic coincidence between natural and historical time. The mathematical and technological operationalization and successful functionalization of the vocabulary of time that until now had served us only for the purposes of self-description indicates only the historical transformability, inner flexibility and contextual boundness even of such highly attuned vocabularies as those of physics, mathematics or logic. The different vocabularies we make use of for differing purposes and in varying contexts are subject to change over time, through which they are respectively related to and distinguished from one another in a varying and contingent way in different historical situations.

The radical temporalization of time expressed in these deliberations had already been outlined in literature by the Austrian novelist Robert Musil. In his novel *The Man without Qualities* he writes, "The train of events is a train unrolling its rails ahead of itself. The river of time is a river sweeping its banks along with it. The traveller moves about on a solid floor between solid walls; but the floor and the walls are being moved along too, imperceptibly, and yet in very lively fashion, by the movements that his fellow-travellers make" (Musil, 1954, p. 174).[4] Within modern philosophy the inner reflexivity of the modern apprehension of time, articulated here by Musil, was founded by Martin Heidegger. In the following considerations the developmental lines highlighted in current theory of time will be set in the context of two basic tendencies that pervade modernity's thinking on time altogether. These basic tendencies of the modern time debate can be described as two ways of temporalizing time (cf. Sandbothe, 1994, 1997). The *objective* temporalization of time in physics contrasts with the *reflexive* temporalization of time in philosophy.

The different ways of temporalizing time appear with particular clarity in the time theories of Martin Heidegger and Ilya Prigogine which form the focus of the present work. Both authors were prominent advocates of pioneering concepts of time in the twentieth century. The philosophical analysis of temporality presented by Heidegger in his early main work *Being and Time* (1927) may be considered the Magna Carta of the philosophy of time in the twentieth century. The Nobel Prize–winning chemico-physical research carried out by Prigogine in the second half of the twentieth century has, from the perspective of thermodynamics, destabilized the time concepts in the physical disciplines of dynamics, quantum theory and cosmology. The present work historically situates Heidegger's and Prigogine's time concepts in the context of the basic tendencies of modern debate on time and uses this basis to relate them to one another systematically. The physical temporalization of time is examined as a historical process taking place at the object level of natural scientific research and culminating in Prigogine's work. The reflexive temporalization of time in philosophy is set alongside the objective temporalization of time in physics as a set of intellectual instruments allowing the objective understanding of time of physics to be critically reinterpreted.

## Notes

1. For a comprehensive bibliography on the subject of time arranged by discipline, see Macey, 1991.

2. I have followed up the influence of electronic media on the scientific concept of time and everyday experience of time in Sandbothe, 1996, 1999. See also Sandbothe and Zimmerli, 1994.

3. There is in general no thoroughgoing correspondence between the German and English editions of the various works of Prigogine referred to by the author in the present work. Wherever possible I have referred to English versions; where, however, this has not proved possible I translate the German. In the case of *Being and Becoming* a revised German edition appeared in 1984. This added a new concluding chapter "Irreversibility and Space-Time Structure"—referred to here—to the previous German edition, which had been based on the English edition of 1980. The 1980 English edition has not been similarly revised [trans.].

4. I am grateful to Wolfgang Welsch for pointing out this quote to me.

PART ONE

# The Objective Temporalization of Time in Physics

To reveal the objective temporalization tendency that took shape in the nineteenth and twentieth centuries with the emergence and development of the physical discipline of thermodynamics I will rely on the reconstruction of the history of the physical understanding of time presented by Ilya Prigogine and Isabelle Stengers in their joint work *Order out of Chaos* (1985). In secondary literature on Prigogine this reconstruction is usually understood to be a merely preliminary work and excluded from considerations.[1] Against this tendency I would like to foreground the connection between Prigogine and Stengers' reconstruction of the immanent pluralization of physical discourse, which took place in the transition from the nineteenth to the twentieth century, and the temporalization of the physical understanding of time that has taken place in the history of modern thermodynamics from Fourier and Carnot, by way of Thomson, Clausius and Boltzmann, through to Gibbs, Onsager and Prigogine.

## The Concept of Reversible Time as the Fundament of Classical Thermodynamics

In the first part of *Order out of Chaos* Prigogine and Stengers begin by setting out the basic features of the concept of reversible time that underlies classical Newtonian physics in order to use this as a demarcational foil for their historical reconstruction of the temporalization of time in modern physics. Under the heading "The Delusion of the Universal" the authors describe the "Triumph of Reason" (Prigogine and Stengers, 1985, pp. 27ff.) as an undertaking deploying mathematical methods and laying claim to global "Identification of the Real" (Prigogine and Stengers, 1985, pp. 57ff.), that is, identification of all domains of reality. In summary, the basic features of the time concept underlying classical physics can be described in terms of the three decisive steps of abstraction which comprise the fundament of modern physics.

The *first*, and fundamental, abstraction and idealization step made by modern science is identified by Prigogine and Stengers as being the reduction in the multitude of different forms of change. According to Aristotle's *Physics*, which remained canonical until the late middle ages, there are "as many types of motion or change as there are of being" (Aristotle, 1984a, Book III, 201a, p. 343). From Galileo onward modern science has limited itself to the basic model of external motion, that is, to change in position. Summarizing critically, Prigogine and Stengers write: "there is only one type of change surviving in dynamics, one 'process,' and that is motion. The qualitative diversity of changes in nature is reduced to the study of the relative displacement of material bodies" (Prigogine and Stengers, 1985, p. 62).

The idea of Newton and Galileo was that all complex forms of change in the broadest sense (μεταβολή, metabolé) could be so far analyzed and divided into their elements that they could be grasped at the microlevel as cases of locomotion (κίνεσις κατὰ τόπο, kínesis katà tópon; (φορά, phorá). This was to include what Aristotle distinguished as the becoming or perishing of a substance (γένεσις καὶ φθορά, génesis kaì phthorá), qualitative change in a substance's properties (ἀλλοίωσισ, ἀlloíosis), as well as processes of increase and decrease (αὔξησις καὶ φθίσις, aüxesis kaì phthísis; Aristotle, 1984a, Book V, Ch. 1, 225a/b, Ch. 2, 226a/b, pp. 380f., 382f.). On this Prigogine and Pahaut write: "Aristotle already understood different modalities of change, above all displacement (local movement) and change in properties (alteration). Classical, as well as relativistic and quantum physics have concentrated on time considered as motion. It seemed as if time, as qualitative change, lay outside their horizon" (Prigogine and Pahaut, 1985, p. 26).

Aristotle's distinctions, however, go beyond the difference between quantitative and qualitative change mentioned by Prigogine and Stengers. Aristotle differentiates the different types of change—μεταβολή—in two respects. According to the first criterion of differentiation—the relation between the start and end points of the change—Aristotle names three types of changes: (κίνησις, γένεσις and φθορά. A process of κίνησις takes place between contrary boundary points. γένεσις and φθορά take place between contradictory boundary points. For its part κίνησις is further divided into three subtypes: quantitative, qualitative, and local change. It is to the subtypes of κίνησις—translated by Barnes as "motion"—that Prigogine and Stengers are referring in the above quote. According to the second criterion of differentiation—the order of the categories—Aristotle distinguishes four forms of μεταβολή: κατὰ τὸ ποιόν, (katà tò poión), κατὰ τὸ ποσόν, (katà tò posón), κατὰ τὸ ποῦ, (katà tò poũ) and κατ' οέσίαν (kat' oésían). Of these, the first three are to be determined as forms of κίνησις (Aristotle, 1984a, Book V, Ch. 1, 225b, p. 381; Conen, 1964, p. 37f.).

In the *second* idealization step carried out by the founders of classical mechanics, locomotion, as the extensionally universalized standard model of change altogether, is at the same time intentionally grasped more narrowly: it is reduced to quantifiable, that is, mathematically calculable, minimal and ideal definitions. Locomotion is no longer understood in the Aristotelian manner as a body's mode of being itself, dependent on its inner striving for its "natural goal" (Aristotle, 1984b, 310b, p. 506; Koyré, 1978, pp. 4ff.) at which it comes to rest. Instead rest and motion are presupposed to be states of equal standing which inertly persist in themselves as long as they are not influenced by external factors.

The restriction of modern physics to spatial motion corresponds to its methodical restriction to the question of the *causa efficiens*, that is, to the efficient cause applied from outside. The transition from the old outlook, based on Aristotle, to the new view becomes particularly clear in the difference between the models with which Kepler, on the one hand, and Newton, on the other, attempted to explain the forces which keep a planet in its orbit. Whereas Kepler still searched for tangential forces which drive the planet onward (*causa finalis*), Newton was interested only in the radial forces which determine the direction of the planet's motion (*causa efficiens*).

At the same time, through the limitation of enquiry to external influences, the question as to "why" the motion as such occurs was no longer even posed; rather only the "how" of the change in motion, that is, the body's acceleration was considered: "Ever since Galileo, one of the central problems of physics has been the description of acceleration. The surprising feature was that the change undergone by the state of motion of a body could be formulated in simple mathematical terms. . . . Galileo discovered that we do not need to ask for the *cause* of a state of motion if the motion is uniform, any more than it is necessary to ask the reason for a state of rest" (Prigogine and Stengers, 1985, p. 57).

Prigogine and Stengers are alluding here to the famous passage in the fourth book of Galileo's *Dialogues Concerning Two New Sciences* (1638) which anticipates the law of inertia that was later to be formulated as the first law of motion by Newton: "Imagine any particle projected along a horizontal plane without friction; then we know . . . that this particle will move along this same plane with a motion which is uniform and perpetual, provided the plane has no limits" (Galileo, 1914, p. 244).

This idealizing presupposition is the precondition for the possibility of mathematical calculation of acceleration, to which Newton then proceeded in a *third* idealization step, one combining the previous two. At the same time, in the "Newtonian Synthesis" (Prigogine and Stengers, 1985, p. 37) the old dichotomy between the forms of motion peculiar to the sublunar and celestial world and to this world respectively was abolished and replaced with a uniform concept of motion, valid for both cosmic and terrestrial conditions: "The formulation of the Newtonian laws of motion made use of two converging developments: one in physics, Kepler's laws for planetary motion and Galileo's laws for freely falling bodies, and the other in mathematics, the formulation of differential or 'infinitesimal' calculus" (Prigogine and Stengers, 1985, p. 57).

Kepler's laws of planetary motion served as a paradigmatic model of calculation for Newton. They were explained by Newton in his *Philosophiae naturalis principia mathematica* (1687) using his formulation of the universal law of gravitation and proven to be compatible with Galileo's laws of falling bodies (Newton, 1968, vol. 2, pp. 206ff.). With this the decisive step was made towards transferring macrocosmic notions of order to the sublunar sphere: "the forces between the planets and those accelerating freely falling bodies are not merely similar, but are the same" (Prigogine and Stengers, 1985, p. 64).

The Newtonian synthesis of Kepler's celestial mechanics with Galileo's laws of free fall prerequired the development of calculus. Only with this was Newton able to succeed in describing the universe as a uniform system, one determined by the fact that "gravitational forces connect any two bodies" (Prigogine and Stengers, 1985, p. 59). Although in the *Principia* (1687) Newton abstains from directly applying the fluxion calculation which he had already developed in *De analysi per aequationes infinitas* (1669) and *Methodus fluxiorum et serierum infinitarum* (1671), the *Principia*'s considerations are, in terms of content, based on the calculus which Leibniz was to develop in the form that remains usual to this day.

The problem originally addressed by calculus is the problem of determining the instantaneous velocity of a body. When one wants to determine the velocity of a moving body at a certain point in time, one is faced by the problem that, by definition, a single point in time encompasses no duration. The instantaneous velocity of a body would accordingly be equal to zero at every moment. In order to solve this paradox—one already formulated by

Zeno—calculus operates by considering limiting values. Instead of the timeless moment which it presupposes, it examines infinitesimally small time intervals whose duration converges to zero. Everything depends on the previous determination of the initial state, that is, on the fixation of the position and velocity of a body at an assumed zero point in time, from which the limit of a series of average velocities over ever decreasing time intervals is extrapolated. The result of this operation is a differential quotient: the first derivative of position with respect to time. Prigogine and Stengers comment: "Instantaneous velocities and accelerations are limiting quantities that measure the ratio between two infinitesimal quantities: the variation of $r$ (or $v$) [that is, of position or velocity—M.S.] during a temporal interval $\Delta t$, and this interval $\Delta t$ when $\Delta t$ tends to zero. Such quantities are 'derivatives with respect to time' " (Prigogine and Stengers, 1985, p. 58).

In order additionally to calculate the acceleration of the body, that is, the rate of change in velocity, brought about by the action of forces (such as gravitation), a further step is necessary: its relationship to the elements of the system acting on it must be considered. The calculation of acceleration is enabled only by the summation of the forces acting on all the system's relevant elements, that is, by integrating over the body. The reciprocal forces acting on elements of the system are thus calculated according to the law of gravitation, which defines the attractive force as being proportional to the masses of the attracting bodies and inversely proportional to the square of their separation. The result of the integration yields the trajectory: a curve containing "all the information acknowledged as relevant by dynamics" (Prigogine and Stengers, 1985, p. 59). From the determination of the current state of a system the trajectory allows the prediction of its future and past with equal exactitude.

The calculation of a dynamic system's temporal change by means of calculus expresses the basic feature of the time concept underlying modern physics. The trajectory is not an arrow marking an irreversible direction, but a line neutral to direction. On this line past and future become interchangeable coordinates which can be calculated equally well from an arbitrarily chosen point. The "denial of the arrow of time" (Prigogine and Stengers, 1993, pp. 9, 42ff.) thus expressed is explained by Prigogine and Stengers when they write in summary:

> In the world of dynamics, change is identified with acceleration or deceleration. The integration of the laws of motion leads to the trajectories that the particles follow. Therefore the laws of change, of time's impact on nature, are expressed in terms of the characteristics of trajectories. The basic characteristics of trajectories

are *lawfulness*, *determinism*, and *reversibility*. . . . In order to calculate a trajectory we need, in addition to our knowledge of the laws of motion, an empirical definition of a single instantaneous state of the system. . . . The remarkable feature is that once the forces are known, any single state is sufficient to define the system completely, not only its future but also its past. At each instant, therefore, everything is given. (Prigogine and Stengers, 1985, p. 60)

The fundamental property of reversibility is hence inscribed in all accounts of change from the start through the preordained structure of the dynamic equations: "The property of reversibility can be quite simply expressed: the dynamic law is such that an (imaginary or computer-simulated) inversion operation $v \rightarrow -v$ of the velocity of each point of the system is equivalent to an inversion operation on the direction of time flow $t \rightarrow -t$" (Prigogine, Stengers, and Pahaut, 1979, p. 44). The determination of reversibility is an attribution which is all but self-evident—yet it was taken to be self-evident by the founders of classical mechanics:

The *reversibility* of a dynamic trajectory was explicitly stated by all the founders of dynamics. For instance, when Galileo or Huygens sought to describe the implications of the equivalence between cause and effect, which they postulated as the basis of their mathematization of motion, they referred to thought experiments such as an elastic ball bouncing on the ground. As the result of its instantaneous velocity inversion, such a body would return to its initial position. (Prigogine and Stengers, 1985, p. 60f.)

For the mathematically represented system, time in this way becomes just that neutral parameter of an "Absolute, True, and Mathematical time" (Newton, 1968, vol. 1, p. 8) which Newton had appointed it to be in his *Principia*: "Absolute, True, and Mathematical time, of itself, and from its own nature, flows equably without regard to any thing external" (Newton, 1968, vol. 1, p. 8). All deviations going beyond the information relevant to dynamics, through which time would acquire a dynamics of its own that would hinder the deterministic ascertainment of the system's past and future, are excluded by Newton. They are disqualified as being aspects of a "Relative . . . Time" (Newton, 1968, vol. 1, p. 8) which is "some sensible and external (whether accurate or unequable) measure of Duration" (Newton, 1968, vol. 1, p. 8) and which "is commonly used instead of True time" (Newton, 1968, vol. 1, p. 8).

In spite of this exclusion of the "commonly used" (Newton, 1968, vol. 1, p. 8) notion of time from the system of Newtonian mechanics, dynamics makes a double claim to universality. Prigogine and Stengers emphasize this when they write: "Newtonian dynamics thus appears to be doubly universal" (Prigogine and Stengers, 1985, p. 59). Firstly, "The definition of the law of gravity . . . contains no reference to any scale of phenomena." And, secondly, "Since gravitational forces connect any two bodies . . . , the only true dynamic system is the universe as a whole" (Prigogine and Stengers, 1985, p. 59).

In his *Opticks* (1704, appendix 1706) Newton sketched the programmatic lines of a future implementation of this universalization in all scientific domains, one which of course has not been redeemable to the present day, and which according to Prigogine and Stengers was destined to failure from the start. In the concluding section he describes the spectrum of applications for his celestial mechanics beyond the inorganic realm of the objects of physics. Thus, with the realm of chemical and biological phenomena in mind, Newton praises the "method of analysis" (Newton, 1952, p. 404) which "proceed[s] from compounds to ingredients" (Newton, 1952, p. 404) as being the universal method of science. The claim expressed in this is clearly highlighted by Prigogine and Stengers early on:

> Newton had no hesitation regarding the universal nature of the laws set out in his *Principia*. Nature is "very consonant and conformable to herself," he asserts in the celebrated Question 31 of his *Opticks*—and this strong and elliptical statement conceals a vast claim: combustion, fermentation, heat, cohesion, magnetism . . . there is no natural process which would not be produced by these active forces—attractions and repulsions—that govern the motion of the stars and that of freely falling bodies. (Prigogine and Stengers, 1985, p. 28)

This universalization of a method gained through orientation towards Kepler's laws of motion, that is, an idealized macrocosmic system, was to become the basic dogma of the classical conception of science: "we have believed in the 'simplicity' of the microscopic" (Prigogine, 1980, p. xiii). In this sense Newton summarizes at the end of the *Opticks*: "All these things being consider'd, it seems probable to me, that God in the beginning form'd matter in solid, massy, hard, impenetrable, moveable Particles . . . ; and that these primitive Particles . . . are incomparably harder than any porous Bodies . . . even so very hard, as never to wear or break in pieces" (Newton, 1952, p. 400). The core of this dogma, which, as Prigogine and Stengers emphasize, is based on a hidden "metaphysical choice" (Prigogine and Stengers, 1988, p. 40), is brought to the point by the two authors in *Order out of Chaos* as follows: "Classical science denied becoming, natural diversity, both considered by Aristotle as attributes of the sublunar, inferior world. In this sense classical science brought heaven to earth" (Prigogine and Stengers, 1985, p. 305).

The three idealization steps reconstructed—the reduction of change to locomotion, the determination of locomotion as a mathematical state according to the law of inertia, and the precise mathematical calculation of this motion by means of calculus—constitute the basic mechanical paradigm of classical physics and simultaneously denote the decisive operations through which the "elimination of time" (Prigogine, 1998, p. 17) is inscribed in the midst of the modern conception of science. It should be noted here that the obliviousness to time in classical physics, critically highlighted by Prigogine and Stengers, was for neither Galileo nor Newton, nor their successors, an explicit or even a methodically reflected procedure.

The first philosophically developed account of time as a parameter (in the sense of modern science) is found with Leibniz in the *New Essays on the Human Understanding* (Leibniz, 1981). In the *Essays*, written by Leibniz in 1704 but first published in 1765, he speaks in the fourteenth chapter of the second book of "the idea of duration . . . , which is a simple and uniform continuum like a straight line" (Leibniz, 1981, p. 152). Leibniz describes its parametric function as a "measure of motion" (Leibniz, 1981, p. 152) in the following way:

> Changes in our perceptions prompt us to think of time, and we measure it by means of uniform changes. But even if nothing in nature were uniform, time could still be determined. . . . Knowing the rules of non-uniform motions, we can always bring them back to comprehensible uniform motions, and by this means predict what will happen through various motions in combination. In this sense time is the measure of motion, i.e. uniform motion is the measure of non-uniform motion." (Leibniz, 1981, p. 152)

The oblivion to the arrow of time contained in the spatializing notion of parameters was dealt with neither by Leibniz nor, later on, by Kant. Only at the close of the nineteenth century—that is, more than two hundred years after Galileo's introduction of the graphical representation of time as a linear dimension in physics—did Boltzmann and Bergson draw scientific attention to the obliviousness of classical physics to time (Prigogine and Stengers, 1993, pp. 42–50; Serres, 1977, pp. 127–142; Szendrei, 1989). Throughout this period the negation of time was carried out, but not noticed—thus itself being negated.

This also holds for d'Alembert, who had nonetheless, in 1754 in the *Encyclopédie*, already pointed out the peculiarity that time appeared in dynamics only as a geometrical parameter. With him this is simultaneously linked with the more refined idea—albeit one which no less detemporalizes time—of a fourth dimension, which was taken up in earnest by Einstein and Minkowsky at the beginning of the twentieth century. D'Alembert writes: "An ingenious man known to me believes that one could, however, regard duration as a fourth *dimension*" (D'Alembert, 1966, p. 1010), where Lagrange is probably meant by the "ingenious man" referred to (Prigogine and Pahaut, 1985, p. 23). But d'Alembert then continues: "This idea can be contested, but it seems to me to have some merit, if only that of being new" (D'Alembert, 1966, p. 1010). D'Alembert, however, did not further pursue the task, suggested in the first sentence, of methodically problematizing the spatial conception of time. Instead he names only the novelty of the idea as an indication of its having "some merit." The "plan to invoke time as a complementary dimension for the description of mechanical phenomena" (Prigogine and Pahaut, 1985, p. 23) first became important for physics with Minkowsky and Einstein. For the great theoreticians of mechanics such as Lagrange and Laplace, time, by contrast, remained "a simple parameter of trajectories" (Prigogine and Pahaut, 1985, p. 23).

That the obliviousness to time is already forgotten when it occurs is a fact not explicitly mentioned by Prigogine and Stengers in *Order out of Chaos*. This fact is first clearly highlighted in later works. Thus in their essay "Evolution and Irreversibility" the authors point out that "the radical character of a negation of all evolution arising from this revolution was first understood with further extension of the theory" (Prigogine and Stengers, 1990, p. 3f.). In the *Paradox of Time* they highlight the dual paradox of classical physics: "Indeed classical dynamics, which was the paradigm of science altogether from the 17th century on, implicitly postulated the radical denial of time. Evidence of the paradoxical character of this denial is that it was never explicitly expressed until Boltzmann's failure. None of the great philosophers, be it Leibniz or Kant, who acknowledged dynamics as the model of human rationality had ever dared to enounce or face up to this implication" (Prigogine and Stengers, 1993, p. 43).

In *Entre le temps et l'éternité* Prigogine and Stengers isolate the fundamental status of the temporal horizon for human experience as the reason for the fact that the obliviousness towards time was itself forgotten: "The idea of a distinction between before and after is . . . a constituent of our experience to such an extent that we cannot describe this [experience—M.S.] without presupposing this difference. . . . Physicists could only read the negation contained in dynamics out of their equations when they were forced to do so by the issue of irreversible processes" (Prigogine and Stengers, 1988,

p. 33f.). It is now time to look at this requirement, which arose with the emergence and scientific establishment of thermodynamics as an independent discipline within the system of physics.

The establishment of the concept of the irreversible arrow of time within physics took place in three steps, corresponding to the three different fields of work in contemporary thermodynamics. The first step is represented by the founding of classical thermodynamics in the nineteenth century, which is oriented towards equilibrium states in ideal and isolated systems. The second and third steps are concerned with the transition from classical to modern thermodynamics which took place in the twentieth century. This does away with the old restriction to the analysis of equilibrium states in ideal and isolated systems. Open systems now come into consideration which, as a result of their exchange of energy and matter with the environment, can no longer attain thermodynamic equilibrium, or which are so distant from the thermodynamic final state that the dynamics of their evolution are no longer directed towards a final state. It is in these systems, which are at the center of the far-from-equilibrium thermodynamics founded by Prigogine, that the variety of different, mutually and complexly linked time forms are first expressed with full precision.

## The Introduction of Irreversible Time in Physics: On the Emergence and Scientific Establishment of Thermodynamics

Thermodynamics developed as a branch of physics from the practical technical interest in the construction and perfection of heat engines. The first effective heat engine was constructed by James Watt in 1782. Because of its close practical orientation and its methods of empirical approximation, thermodynamics was for a long time treated as a kind of stepchild of physics. Because of the degree of complexity of its object, thermodynamics abstains from the outset from representing thermodynamic systems microphysically. The enormous number of particles comprising the systems examined in thermodynamics is made clear by Prigogine and Stengers: "Physical chemistry often employs Avogadro's number—that is, the number of molecules in a 'mole' of matter (a mole of matter always contains the same number of particles, the number of atoms contained in one gram of hydrogen). This number is of the order $6 \times 10^{23}$, and it is the characteristic order of magnitude of the number of particles forming systems governed by the laws of classical thermodynamics" (Prigogine and Stengers, 1985, p. 120, footnote). In taking account of the complexity of its object, thermodynamics works exclusively with macroscopic parameters like pressure, volume and temperature. Yet what was once considered "the 'failing' of a physics which seemed condemned to empirical approximations and predictions has now," Prigogine and Stengers claim, "generally become the key to a renewal of our notions of matter, space and time in physical theory" (Prigogine and Stengers, 1981, p. iii).

The beginnings of this new estimation of the significance of thermodynamics for twentieth century physics can already be found with the British physicist Arthur Eddington. In his book *The Nature of the Physical World*, which Prigogine and Stengers explicitly refer to in the introduction to *Order out of Chaos*, he writes: "from the point of view of philosophy of science the conception associated with entropy must . . . be ranked as the great contribution of the nineteenth century to scientific thought" (Eddington, 1943, p. 103). It was also Eddington who recognized early on the particular significance of thermodynamics as lying in the introduction of "time's arrow" (Eddington, 1943, p. 68), which had been suppressed by the classical conception of science, to research. Hence Eddington explicitly distinguishes the second law of thermodynamics from the classical "primary law, . . . indifferent to a time-direction" (Eddington, 1943, p. 76) by designating it an independent "secondary law" (Eddington, 1943, p. 76) "which recognises a distinction between past and future more profound than the difference of plus and minus" (Eddington, 1943, p. 66f.). And at the end of the chapter quoted Eddington summarizes: "I am interested in entropy not only because it shortens calculations which can be made by other methods, but because it determines an orientation which cannot be found by other methods" (Eddington, 1943, p. 109).

First order thermodynamics was founded in the nineteenth century with three steps, made by Sadi Carnot, William Thomson and Rudolf Clausius, and Ludwig Boltzmann. The third step, made by Boltzmann, assumes particular importance both for the scientific establishment of thermodynamics and for discussion about the time-theoretical presuppositions of modern physics. To begin with, however, the basis laid by Carnot, Thomson and Clausius in the first two foundational steps will be dealt with.

## The Emergence of Thermodynamics in the Nineteenth Century

The prehistory of the emergence of thermodynamics begins with Fourier's discovery of the law of conduction against the background of the discovery of the principle of conservation which was taking place in the first half of the nineteenth century. The first preparatory step towards the development of thermodynamics—decades before the concept was coined and research into heat became the independent object of an exact science—triggered intense disputes in the scientific world already at the beginning of the nineteenth century. Fourier's discovery of the law of conduction, which was to become "the starting point of an investigation into the nature of irreversibility" (Prigogine and Stengers, 1985, p. 105), led already in the first third of the nineteenth century to a decisive break with the self-apprehension of the classical conception of science dating back to Newton. This was a break which was to grow into a deep-reaching crisis in the course of the nineteenth and twentieth centuries and which Newton's proponents and heirs sought to patch up from the first and with continually changing strategies over the course of time.

Although Fourier had already been investigating the propagation of heat in solid bodies from 1807 onwards, his referees Lagrange, Laplace and Legendre, as strict proponents of Newtonian doctrine, ensured that only in 1822 was it first possible for Fourier to summarize and publish his results in the book *The Analytical Theory of Heat*. What was it about Fourier's work that was threatening for the protagonists of classical mechanics?

Prigogine and Stengers describe Fourier's discovery as follows: "Fourier's law, when applied to an isolated body with an unhomogeneous temperature distribution, describes the gradual onset of thermal equilibrium. The effect of heat propagation is to equalize progressively the distribution of temperature until homogeneity is reached" (Prigogine and Stengers, 1985, p. 105). This fact, according to which heat propagation takes place as an irreversible process moving towards a homogeneous final state, stood in diametrical opposition to the reversibility concept of Newtonian mechanics. What was provocative for contemporary science of this time was, above all, that the irreversible process of conduction could be simply and elegantly represented mathematically. The flow of heat is proportional to the temperature gradient—that is Fourier's simple equation.

In his *Course in Positive Philosophy* (1830–1842) Auguste Comte immediately drew the consequence from this result and ennobled Fourier's result as one of the first purely positivist research results in modern physics (Prigogine and Stengers, 1985, p. 104f.; Serres, 1974, pp. 159–185; Serres, 1975; Serres, 1977, esp. pp. 134–137). Comte highlighted here that Fourier's "theory of the distribution of heat" (Comte, 1974, p. 196) was to be considered an example of the direct "application . . . of mathematical analysis" (Comte, 1974, p. 195), one not mediated by preordained laws. According to Comte, Fourier succeeded by bringing "his subject up to such a point of positivity as to place it next to the study of gravity" (Comte, 1974, p. 204); this high degree of positivity was seen to be directly employed "when we can seize the fundamental law of phenomena, so as to make it the basis of a series of analytical deductions" (Comte, 1974, p. 196). With a view to Descartes, Kepler and other physicists "even long after Galileo's time" (Comte, 1974, p. 203) Comte criticizes the formation of hypotheses that aim at "determining the general agents to which different kinds of natural effects may be referred" (Comte, 1974, p. 200). He views this way of explaining and justifying as being a last influence of metaphysical philosophy and, against this, Comte continues, the "labors of Fourier will evidently free thermology . . . from all fantastic notions about imaginary fluids" (Comte, 1974, p. 204) and ought to be considered an example of the positivist principle of physics, which states "that every scientific hypothesis . . . must relate exclusively to the laws of phenomena, and never to their mode of production" (Comte, 1974, p. 204).

And Comte went a step further still. He declared heat, alongside gravitation, to be an independent universal of physical research and made this distinction the basis of his classification of the sciences: "To the mechanical equilibrium between forces the positivist classification simply adds the concept of thermal equilibrium" (Prigogine and Stengers, 1985, p. 105). Indeed in his positive philosophy Comte divides physics into the domains of organic and inorganic physics. Inorganic physics is for its part divided into astronomy and terrestrial physics (Comte, 1974, p. 44f.). The latter consists once again of "barology" (Comte, 1974, Book III, Chapter 2), or the theory of weight, and "thermology" (Comte, 1974, Book III, Chapter 3), or the theory of heat.

With this last classification, one introduced by him, Comte is explicitly opposing attempts to subordinate heat—say as thermal matter—to the theory of massive bodies (Comte, 1974, p. 201f.). Instead he highlights the universality and independence of weight and heat: "gravity manifests itself in the same way in all bodies; and the same with phenomena of heat" (Comte, 1974, p. 193). Serres writes of Comte: "He recognizes the tear and sees that physics and science in general can no longer be uniform: that there are Newton *and* Fourier, that the two universals cannot be reduced to one another, that the sun is the focus of our world-system's ellipse and simultaneously the focus of heat, and that different things are hidden behind these two similar words. And that, hence, the unity, which was provided by nature in the age of metaphysics and once by God, has been forever lost. He admits that there are several laws" (Serres, 1977, p. 136). Comte diagnoses the first profound foundational crisis of modern physics.

At the same time, in terms of the philosophy of science, the account of the internal pluralization of physics, which finds expression in Comte's emphasis of the equality of mechanics and thermodynamics, should be relativized. Comte's classification has a hierarchical background which Prigogine and Stengers suppress in their account of Comte. Although the theory of weight and the theory of heat are, according to Comte, not reducible to one another, their relationship to one another nonetheless remains determined by the scientific priority of mechanics, which for Comte continues to come first among the departments of physics (Comte, 1974, p. 204). The theory of heat is subsequent to this in terms of its degree of scientific universality: "First will come Weight . . . , regarded statically and dynamically [i.e. as mechanics—M.S.] . . . ; weight being absolutely universal" (Comte, 1974, p. 204). On this Serres, who clearly brings out Comte's inconsistency, writes: "The antagonism demands a choice. Comte chooses. He leaves the order of the world and of science intact, conserves classical history and the monument to Lagrange, and plugs the hole opened by the antagonism by assigning Fourier a place outside of the text—in the dedication" (Serres, 1977, p. 136).

Similar ambivalence is found with Eddington. Just as Comte in relation to Fourier's law of conduction, Eddington attributes the second law the status of a natural law, yet with the graduation of first and second laws simultaneously introduces the prescription of a hierarchy which disqualifies the second law in contrast to the first, actual, natural law: "the second law of thermodynamics . . . stands aloof from all the rest. But this law has no application to the behaviour of a single individual, and . . . its subject matter is the random element in a crowd" (Eddington, 1943, p. 66f.). In doing this Eddington even explicitly points out the hierarchical structure of his distinction: "I have called the laws controlling the behaviour of single individuals "primary laws," implying that the second law of thermodynamics, although a recognised law of Nature, is in some sense a secondary law" (Eddington, 1943, p. 75).

The fact which is of primary concern to Prigogine and Stengers remains unaffected by these limitations: the accentuated consideration of the theory of heat in Comte's classification makes it strikingly clear that the universalism of Newtonian mechanics, which had just been brought to fruition by Laplace, sustained a violent recoil within the lifetime of its protagonist, a blow which was immediately capitalized upon by Comte's theory of science: "A physical theory had been created that was every bit as mathematically rigorous as the mechanical laws of motion but that remained completely alien to the Newtonian world" (Prigogine and Stengers, 1985, p. 104; cf. also Comte, 1974, p. 108). In a later passage, with the scandal brought about by Fourier and brought to a head by Comte's philosophy of science in mind, Prigogine and Stengers write: "The identity of the mathematical formulation of natural laws and of classical mechanics was broken forever" (Prigogine and Stengers, 1981, p. 282). In this spirit Comte emphasizes that "there is something . . . chimerical in attempts at universal explanation by a single law. . . . Our intellectual resources are too narrow, and the universe is too complex, to leave any hope that it will ever be within our power to carry scientific perfection to its last degree of simplicity" (Comte, 1974, p. 37).

The internal pluralization of physics which came to light in the science and philosophy of science of the nineteenth century supports Prigogine and Stengers' initial thesis of the openness, in principle, of physical thinking for what eluded its own classical conception, a conception which, particularly in the nineteenth century, was flourishing (Prigogine and Stengers, 1985, pp. 1–26). At the height of its success a kind of counter-discourse was beginning to inscribe itself within the discourse of mathematical natural science; even *before* the foundational crisis, a profound foundational crisis was taking shape here which has to date scarcely been adequately considered in the history of science.

The other important line of development, which was to nourish the theoretical thermodynamics formulated by Thomson and Clausius in the 1850s and 1860s, had—unlike Fourier's discovery—a more stabilizing effect on the program of classical mechanics (Moscovici, 1977, pp. 400–406). The principle of energy conservation introduced a new "unifying element" (Prigogine and Stengers, 1985, p. 108) into nineteenth-century science, with which a "new golden age in physics began to take shape, an age that would lead to the ultimate generalization of mechanics" (Prigogine and Stengers, 1985, p. 111).

In his essay "Energy Conservation As an Example of Simultaneous Discovery" (Kuhn, 1969), from which Prigogine and Stengers' account draws support, Thomas S. Kuhn shows how a new "look" (Kuhn, 1969, p. 324) to the physical sciences developed between 1800 and 1835, one documented in popularized form in Mary Sommerville's book *On the Connexion of the Physical Sciences* (Sommerville, 1975). The impression articulated by Sommerville of a new connection between the different disciplines of physics is based on a chain of discoveries following Volta's invention of the battery in 1800. Common to the discoveries of Volta, Oersted, Seebeck, Peltier, Faraday et al. which Kuhn, Prigogine and Stengers list is that their respective objects were mutually independent transformation processes between "Chemical Affinity, Electricity, Heat, Magnetism and other powers of Matter" (Faraday, quoted from Kuhn, 1969, p. 327; cf. Jones, 1870, vol. 2, p. 47). In this context Prigogine and Stengers speak of a "network [of new processes] that ultimately linked all the new fields of physics with other, more traditional branches, such as mechanics" (Prigogine and Stengers, 1985, p. 107).

The principle of the conservation of energy, which was formulated in parallel by many different scientists (Mayer, Joule, Colding, Helmholtz, Grove, Faraday et al.) in the 1840s, proves, according to Kuhn, to be the "theoretical counterpart" (Kuhn, 1969, p. 325) to these initially isolated transformation processes. The transition from "the earlier network of laboratory conversion processes" (Kuhn, 1969, p. 325) to the theoretical formulation of the law of energy conservation did not, however, take place in the sense of a cumulative and logically strict growth in knowledge. Rather, Kuhn emphasizes: "This realization came neither all at once, nor fully to all, nor with complete logical rigor" (Kuhn, 1969, p. 328).

In an endnote Kuhn ekes out the "logical shortcoming" (Kuhn, 1969, p. 346, note 26) which attaches to the derivation of the conservation principle from the empirical system of the manifold transformability of energy. Proceeding from the fact that every force can bring about all others, and via the intermediate step that this prerequires the equality of cause and effect, the pioneers of the conservation principle concluded the impossibility of energy loss in transformations, that is, the necessary conservation of energy throughout all transformation processes. The logical problem with this conclusion lies, according to Kuhn, in the more detailed justification of the intermediate step. This justification states that the equality of cause and effect refers to a comprehensive quantitative equivalence, since in the case of a quantitative asymmetry between cause and effect a suitably chosen sequence of transformations would lead to the generation of energy, which, however, already had to be excluded by Lazare Carnot's proof of the impossibility of a *perpetuum mobile*. Kuhn's criticism refers to this intermediate step: "Strictly speaking, this derivation is valid only if all the transformations of energy are reversible, which they are not" (Kuhn, 1969, p. 346, note 26). For there is a characteristic asymmetry in the system of transformations which, in spite of the quantitative equivalence of cause and effect, destroys the reversibility of the processes, and hence the implicitly presupposed reversibility of the cause-effect relationship.

With this critical footnote Kuhn identifies the gap in the derivation of the conservation principle which, later in the scientific history of the nineteenth century, made it necessary to place the second law of thermodynamics alongside the principle of energy conservation, which was later to be formalized as the first law of thermodynamics. The second law makes this macroscopic irreversibility mathematically determinable by introducing the entropy concept. In the history of science the sequence of first and second laws reads in chronological opposition to the subsequent numbering. The law we today call the second law was the first to be discovered. This historical priority, which for Prigogine and Stengers is simultaneously linked with a material priority, is highlighted by the authors under the heading "Heat Engines and the Arrow of Time" at the beginning of the third section of Chapter IV of *Order out of Chaos*: "The original formulation of the second law of thermodynamics, which would lead to the first quantitative expression of irreversibility, was made by Sadi Carnot in 1824, before the general formulation of the principle of conservation of energy by Mayer (1842) and Helmholtz (1847)" (Prigogine and Stengers, 1985, p. 111).

As a technician concerned with the calculation and optimization of the efficiency of steam engines, Carnot was less influenced by the "new look" science at the beginning of the nineteenth century than were the pioneers of the conservation principle. Unlike those of his colleagues who occupied themselves empirically with a multitude of various transformation processes (such as, above all, Faraday and Grove), or who took the "notion of an underlying imperishable metaphysical force" (Kuhn, 1969, p. 336) as their starting point, or adduced such a force in the explanation of relationships taking shape empirically (as, for example, did Mayer and Helmholtz), Carnot—on the basis of his practical technical interest—concentrated solely on the transformation of heat into work in the steam engine.

Carnot begins his *Reflexions* with the following far-sighted reflection on the practical importance of the steam engine: "The study of these engines is of the utmost interest. Their importance is immense and their use is increasing daily. They seem destined to bring about a great revolution in the civilized world. . . . It seems that one day it [the heat engine] must become a universal source of power and in this respect supplant animals, water and wind" (Carnot, 1986, p. 61). Neither the many practical suggestions resonating throughout Carnot's work, nor his sensitivity for the profound consequences of technological development have yet to be sufficiently appreciated.

It was the technical, pragmatic approach, one less burdened by contemporary developments in science, which led Carnot and his successors to insights that could not be explained by the universal equivalences emphasized by the conservation principle. The conservation principle ignores the fact that, although energy remains quantitatively unchanged in transformation processes, irreversible changes in quality can nonetheless simultaneously occur which lead to transformation events being non-reversible in spite of energy conservation. It is this characteristic irreversibility, one expressing the temporal asymmetry excluded and ignored by the founders of the conversation principle, which was already pointed out by Fourier.

In just this asymmetry, however, according to Prigogine and Stengers were manifested "the last remnants of the spontaneous and intrinsic *activity* displayed by nature" (Prigogine and Stengers, 1985, p. 120) and, with this, the enduringly merely partial controllability of a nature that resists classical mechanics' universal notions of control. For this reason it was suppressed by the advocates of classical mechanics. Prigogine and Stengers write: "The power of nature is thus concealed by the use of equivalences. However, there is another aspect of nature that involves the boilers of steam engines, chemical transformations, life and death, and that goes beyond equivalences and conservation of energy. Here we reach the most original contribution of thermodynamics, the concept of irreversibility" (Prigogine and Stengers, 1985, p. 111).

In 1824 Carnot was not yet familiar with the conservation principle and thus could not relate his work to it. This he first did later in notes, begun after the publication of his *Reflexions* (1824) and continued until his death (1832), in which he proceeded to an independent formulation of the conservation hypothesis (Carnot, 1986, pp. 160ff.). Unlike Clausius and Thomson, however, Carnot assumed that the conservation hypothesis was fully incompatible with the main thesis of his *Reflexions*. Thus he himself did not recognize the corrective significance of his early research with regard to energy conservation.

His pioneering considerations, informed by the model of the water wheel examined by his father Lazare Carnot, built upon some anachronistic ideas which, whilst retaining Carnot's results, were corrected by Clausius and Thomson. Unlike Joule, Clausius and Thomson, who were able to verify and establish in physics the kinetic theory of heat anticipated by Rumford and Davy, Carnot set out from the caloric theory developed by Black (Elkana, 1974, pp. 60ff.). This theory allowed Carnot to understand the steam engine by analogy to the water wheel. According to this idea, work in the steam engine, as in the water wheel, arises through the flow of caloric, imagined to be a weightless fluid, from the hot steam-boiler to the cool condenser. In overcoming the temperature gradient heat, according to Carnot's second false assumption, is conserved. Carnot had not yet grasped the difference between heat and energy. Yet, in spite of these false presuppositions, the equations formulated by Carnot were factually correct.

Carnot's achievement lies in having recognized that the condition of possibility for every transition of heat and work is as great a difference in temperature as possible between the two heat sources to be integrated in the Carnot engine. The Carnot engine is not a real engine, but a model summarizing the basic conditions and examining the processes fundamental to all real heat engines in idealized laboratory conditions. Using this engine Carnot showed that without a temperature gradient no work could be obtained. The second heat source is necessary here because, following the first cycle of the Carnot process, heat must be removed to return the system to its initial state and to begin the cycle once again. It is this energy, which must be removed as heat (through dissipation), that is responsible for the fact that the efficiency of every conceivable heat engine is much less than 100 percent. It was Carnot who first recognized this and thus pointed out the specific asymmetry which means that although work can be converted completely into heat, the reverse process must take account of the inevitable and irreversible dissipation of energy which cannot be converted into work but is emitted as heat and hence dispersed.

In Carnot's eyes there was nothing threatening or provocative about this dispersion of energy. It was a technical problem, to be addressed with technical means, but one which for Carnot had no farther-reaching physical implications. This purely technical view was precipitated in Carnot's definition of dissipation. For Carnot the dispersion of energy was solely the effect of the necessary exchange of heat between the two heat sources in which the heat reservoirs were assumed to be infinite so as to guarantee the constancy of the temperature difference. The dispersion of energy was a temporary phenomenon for Carnot with no implications at all for the total energy stock. Accordingly, internal dissipation—which already takes place within the system as a result of friction and conduction, independently of the exchange between the two heat sources—remained excluded from Carnot's idealization. Carnot's solution to the problem rests on this double abstraction.

Carnot does not attempt to avoid loss through dissipation. The unavoidability of heat loss is assumed in the construction of the engine. The strategy of its constructor was far more to resolve, through cunning analysis, the entire irreversible process into infinitesimally small and (in the ideal conditions of the double abstraction mentioned above) quasi-reversible single steps in order to make the irreversible event controllable within the artificially closed Carnot cycle subsystem, whose energy balance is abstracted from the balance of the surrounding total system.

In his calculations Carnot was concerned with "the effect of the combustion, which permits the maintenance of the temperature difference between the two sources" (Prigogine and Stengers, 1985, p. 114). For this reason the two heat sources included by Carnot's apparatus were assumed to be unlimited reservoirs and did not contribute to the loss calculation. Similarly excluded by Carnot were all losses through friction, impacts, air resistance etc.

For only "when defined in terms of its reversible transformations [can] the *thermodynamic object* . . . be controlled through its boundary conditions: any system in thermodynamic equilibrium whose temperature, volume, or pressure are *gradually* changed passes through a series of equilibrium states, and any reversal of the manipulation leads to a return to its initial state" (Prigogine and Stengers, 1985, p. 120).

The uncontrollable component of irreversibility, which first appears at the level of the total energy balance by incorporating the factual finiteness of the heat sources and by considering internal losses through friction and conduction, was acknowledged by Carnot only in his exclusion of them. For this reason Prigogine and Stengers summarize: "Looking back it becomes clearer what Carnot achieved by founding thermodynamics and silencing the heat boiler. By assuming *two sources* he separated what could be idealized and was linked with reversible transformations from that which is essentially irreversible within an engine, namely the combustion process which generates the motion" (Prigogine and Stengers, 1981, p. 128).

William Thomson (Lord Kelvin) was the first to do away with this parenthesis. Unlike Rudolf Clausius, who had already published a essay in 1850 that led the way for the further development of Carnot's considerations and for the mathematical foundation of thermodynamics, Thomson, in his works of 1851 and 1852, accentuated the aspects of the irreversibility of energy dissipation and the finiteness of the reservoirs. Hence from early on Thomson concentrated on those aspects through which thermodynamic processes differed from the reversible motions assumed by mechanics. Thus he was the first to look at the real heat losses which go far beyond the losses considered by the idealized model. Prigogine and Stengers write: "the question raised by Carnot and Clausius led to a description of ideal engines that was based on conservation and compensation. In addition, it provided an opportunity for presenting new problems, such as the dissipation of energy. William Thomson, who had great respect for Fourier's work, was quick to grasp the importance of the problem, and in 1852 he was the first to formulate the second law of thermodynamics" (Prigogine and Stengers, 1985, p. 115).

Thomson's priority in the discovery of the second law, claimed here by Prigogine and Stengers, is disputed. Coveney and Highfield, for example, highlight by contrast the priority of Clausius: "Clausius recognised, dimly at first, that this meant that heat loss was irreversible. . . . His breakthrough in 1850 fully justifies naming Clausius as the scientific father of the arrow of time, but at that stage his ideas were rather ill-defined" (Coveney and Highfield, 1990, p. 150). Sambursky's evaluation, which similarly highlights Clausius's importance for the question of time, seems to point in the same direction: "Clausius's formulation of the principle that the sum of entropies for a closed system always grows, or at least does not decrease, immediately suggests the problem of the passage of time from which mechanics remained completely untouched" (Sambursky, 1978, p. 41f.). Clausius's formulation of the *second law* using the entropy concept—as Sambursky points out elsewhere (Sambursky, 1978, p. 479)—stems, however, from the year 1865. This was the year in which Clausius introduced the concept of entropy in his work "On different, more easily applied, forms of the main equations of mechanical thermodynamics" (Clausius, 1865, p. 46).

In quite the same spirit as Coveney and Highfield, Segrè highlights: "Carnot had created thermodynamics by initiating its powerful methods and discovering its second principle. Mayer, Joule, and Helmholtz had formulated the first principle. The synthesis of both is due to Clausius and, later, to Thomson" (Segrè, 1984, p. 227). And Segrè continues: "Clausius was the first to see clearly through the apparent contradictions of the then-prevalent theories of heat and to give a systematic treatment of the new science" (Segrè, 1984, p. 228). For support Segrè draws upon Gibbs, who, in his textbook *Elementary Principles in Statistical Mechanics* (Gibbs, 1902), had identified Clausius's treatise "On the Motive Power of Heat, and on the Laws which can be deduced from it for the Theory of Heat" (Clausius, 1921, original 1850) as being the foundational document of thermodynamics as an exact science. This estimation was also shared by Ernst Mach who, with Clausius's 1850 essay in mind, writes: "Clausius first perceived that we can assume, with Carnot, the dependence of the performance of work on the quantity of heat transferred without having to give up the principle of Mayer and Joule of the equivalence of heat and work" (Mach, 1986, p. 254).

For a long time in the English-speaking history of science it was usual to neglect Clausius's achievements in a factually unfounded manner, since his works were not sufficiently well known. Clausius himself had already pointed out this deficiency to Maxwell, who in the 1871 first edition of his *Theory of Heat* (Maxwell, 1871)—without even discussing Clausius—ascribed priority in the discovery of the second law to his compatriot Thomson (Clausius, 1887, p. 360f.). By contrast Prigogine and Stengers have good reason to advocate the priority of Thomson. Prigogine and Stengers (unlike Gibbs and Segrè) do not apprehend the second law simply as a synthesis of Carnot's recognitions and the conservation principle. For them the interesting aspect for the theory of time lies precisely in the confrontation and tension prevalent between the first two laws of thermodynamics. This tension is identified by Moscovici when he writes: "This theoretical result, which makes natural order appear as a series of transformations with a direction and manifesting an evolution, constituted a promise and a scandal" (Moscovici, 1977, p. 405). And he continues: "While on the one side one attempted—admittedly without much luck—to fulfil the programme which had been revived by the principle of energy conservation, on the other side one was quick to attenuate the scandal triggered by the second principle, that of energy loss, and by the directionality etched in the transformation of natural forces" (Moscovici, 1977, p. 406).

Indeed Clausius, in 1850, was the first to notice the necessity of adding a second physical law alongside the conservation principle. But whereas, in his notebooks, Carnot had highlighted the fundamental incompatibility of his early work with the conservation principle and thought it necessary to direct this against his own research, Clausius in 1850 accentuated the fundamental agreement of the new physical law and the conservation principle, even, and precisely, with regard to the reversibility of the transformation processes that were occurring. This is also pointed out by Brush when he emphasizes that in his 1850 considerations Clausius "has not yet grasped the irreversibility implications of Carnot's theory" (Brush, 1976, vol. 2, p. 569). Brush hence comes to the conclusion: "While there is some justification for the usual view that this paper contains a complete formulation of the rules of thermodynamics, Clausius has not yet put together the pieces of the complete Second Law as we now know it" (Brush, 1976, vol. 2, p. 570).

This also becomes clear in that for Clausius the first law, and the universal equivalence and reversibility entrenched within it, remains prior in the order of foundation, that is, the primary and fundamental law. Thomson differs. As early as in 1852 he generalized the second law as a cosmic law (Thomson, 1882b)[2] which, although it can be linked with conservation, opposes and claims priority over the conservation principle with the fundamental irreversibility of the transformations it describes. It is this double aspect—the sharp accentuation and the cosmological universalization of irreversibility—which Prigogine and Stengers adduce as justification of Thomson's priority over Clausius.

Although not pointed out by Prigogine and Stengers themselves, the accentuation of irreversibility is, in substance, already found in Thomson's 1851 essay "On the Dynamical Theory of Heat" (Thomson, 1914), in which Thomson was the first to highlight the phenomenon, discovered by Fourier, of heat propagation as the cause of the factual loss of power in steam engines. By taking account of dissipation, the dispersion of energy distinguishing real thermodynamic transformation processes from Carnot's ideal model is incorporated in the energy balance. So with real heat engines the heat source is always finite (Thomson, 1914, section II, pp. 20ff.). Whereas Carnot's and

Clausius's model of ideal heat engines "does not mention the irreversible processes that are at the basis of its realization" (Prigogine and Stengers, 1985, p. 114), Thomson starts to include the oven in which the coals combust in the balance.

It was also Thomson who introduced the term "thermodynamics" to designate the new science of heat. This, however, took place not—as Atkins claims (Atkins, 1984, p. 4)—in the 1851 essay, but—as Truesdell makes clear (Truesdell, 1980, p. 168)—already in the 1849 work "An Account of Carnot's Theory of the Motive Power of Heat; with Numerical Results Deduced from Regnault's Experiments on Steam" (Thomson, 1882a). The concept of thermodynamics is not mentioned in Thomson's 1851 essay. Alongside its significance in terms of content, the 1851 essay has another significant merit in terms of conceptual history: here Carnot's discovery is explicitly designated the "second law" for the first time (Brush, 1976, vol. 2, p. 570f.). It is only in the expanded version of this essay, published by Thomson in 1854, that the term "thermodynamics" is found again, now appearing even in the title.

Nonetheless there is good warrant for Prigogine and Stengers' presenting not Thomson's 1851 essay, but the text of 1852 as being the foundational document of thermodynamics. An explicit and general statement about the fact of irreversibility is first found in the 1852 text, which in connection with this carries out the universalization at the same time. This coupling, however, leads Brush to focus completely on the aspect of universality in his history of thermodynamics: "Although one can find scattered statements in the technical literature before 1850 to the effect that something is always lost or dissipated when heat is used to produce mechanical work, it was not until 1852 that William Thomson . . . asserted the existence of 'A universal tendency in nature to the dissipation of mechanical energy' " (Brush, 1966, p. 8).

By over-accentuating the universality aspect Brush loses sight of what Prigogine and Stengers view as the decisive basic difference between Thomson and Clausius. In his analysis of ideal heat engines, Clausius—unlike Thomson in his works of 1851 and 1852—had neglected the real losses which occur due to the dissipation of energy between heat sources and which lead to the unattainability for real engines of the ideal efficiency predicted by theory. With regard to Clausius's early essay "On the Motive Power of Heat" (Clausius, 1921) of 1850 Prigogine and Stengers highlight that its author "was no more concerned than Carnot with the losses whereby all real engines have an efficiency lower than the ideal value predicted by the theory.

His description, like that of Carnot, corresponds to an idealization" (Prigogine and Stengers, 1985, p. 114).

By contrast already in his 1854 essay (Clausius, 1864), as well as in his later work from the year 1865 (Clausius, 1865), Clausius highlighted and made quantifiable the factual dissipation of energy through which the irreversibility of processes can be explained. In a later passage Prigogine and Stengers write:

> In 1865 . . . Clausius . . . introduced a new concept, *entropy*. His first goal was to distinguish clearly between the concepts of conservation and of reversibility. Unlike mechanical transformations, where reversibility and conservation coincide, a physiochemical transformation may conserve energy even though it cannot be reversed. This is true, for instance, in the case of friction, in which motion is converted into heat, or in the case of heat conduction as it was described by Fourier. (Prigogine and Stengers, 1985, p. 117)

Accordingly the task of the function of state—entropy S—introduced by Clausius was to express mathematically "the distinction between 'useful' exchanges of energy in the Carnot cycle and 'dissipated' energy that is irreversibly wasted" (Prigogine and Stengers, 1985, p. 117) that was now also being addressed.

Nonetheless in his later works Clausius continued to focus on the aspect of conservation by relating the concept of entropy primarily to the exchange of energy within the ideal cycle (which had been made reversible using the methods of calculus). This exchange within the cycle occurs between the two heat sources. Hence Prigogine and Stengers designate this entropy flow $d_eS$, in order to make clear Clausius's renewed abstraction of dividing the total entropy into two parts. This external entropy flow contrasts with the internal entropy flow $d_iS$, which occurs not with the exchange of heat between the two sources, but within the system itself. Prigogine and Stengers explain: "the notations $d_eS$ and $d_iS$ have been chosen to remind the reader that the first term refers to *exchanges (e)* with the outside world, while the second refers to the irreversible processes *inside (i)* the system" (Prigogine and Stengers, 1985, p. 118). For ideal engines it suffices to consider $d_eS$ alone, for real engines the total entropy is comprised of $d_eS$ and $d_iS$ together.

In his 1865 work Clausius "was able to express quantatively the entropy flow $d_eS$ in terms of the heat received (or given up) by the system. In a world dominated by the concepts of reversibility and conservation, this was his main concern. Regarding the irreversible processes involved in entropy production, he merely stated the existence of the inequality $d_iS/dt > 0$" (Prigogine and Stengers, 1985, p. 119). In this simple inequality, which Clausius did not further expand on in his work, is concealed the fundamental irreversiblity which distinguishes the (necessarily positive, i.e. directed) change in internal entropy from the (either positive or negative, i.e. reversible) change which the system experiences through external entropy flows. Clausius was interested only in reversible and hence controllable changes. His endeavor was to separate the irreversible and hence uncontrollable processes as far as possible from the reversible ones and to exclude them from his calculations.

Thus at the very beginning of his 1865 treatise Clausius highlights "that . . . in the following, until it is explicitly stated that non-reversible changes are also to be included in the examination, it [might] always be assumed that we are concerned only with reversible changes" (Clausius, 1865, p. 8). Only at the end do we read: "All the preceding observations referred to changes which take place in a reversible manner. We now want to draw the *non-reversible changes* into the circle of observations, in order to pursue at least briefly the main issue of how they are to be treated" (Clausius, 1865, p. 40). Yet in the mathematical analysis of those changes "which do not comprise a cyclic process, but by which it [the body] attains a final state differing from the initial state" Clausius is then concerned precisely to "make a cyclic process of them retrospectively" (Clausius, 1865, p. 54).

It is exactly these irreversible processes, however, which were the object of Thomson's cosmological conception and his prioritization of the second law. Prigogine and Stengers comment: "Since Thomson's cosmology is not merely a reflection of the new ideal heat engine but also incorporates the consequences of the irreversible propagation of heat in a world in which energy is conserved. This world is described as an engine in which heat is converted into motion only at the price of some irreversible waste and useless dissipation" (Prigogine and Stengers, 1985, p. 115). On the basis of Thomson's cosmological universalization of the second law in 1852, Helmholtz developed his theory of the universe's "heat death" (Helmholtz, 1884, p. 43; Helmholtz, 1962, p. 74) two years later. With this Thomson and Helmholtz transferred the fact, implicitly contained in Thomson's 1851 work, that "unlike dynamic objects, thermodynamic objects can only be *partially* controlled" (Prigogine and Stengers, 1985, p. 120) on the cosmic scale.

The difference in content between Thomson and Clausius expressed in this, one decisive for Prigogine and Stengers, remains unnoticed by Coveney and Highfield as well as by Gibbs and Mach. The compromise solution suggested by Cardwell is also unsatisfactory for the same reason. This suggested solution attempts to resolve the dispute over priority by declaring the development of the second law of thermodynamics to have been a "simultaneous discovery" (Cardwell, 1971, p. 288). This solution is only meaningful when it considers that it is not one and the same thing which was being simultaneously discovered, but that the problem provided by Carnot was picked up in parallel by two authors in a different way and answered with differently accented solutions.

Of interest in this context is a comment of Mach's, who in his summary initially emphasizes that the "contributions to thermodynamics of Thomson and Clausius must be regarded as of equal importance" (Mach, 1986, p. 280f.), but then notes a difference on the level of style between Thomson and Clausius:

> However, with respect to the form of the exposition, there is an important distinction between them. Thomson's exposition is always quite frank about the difficulties which he met, the paths followed by him are always the shortest and simplest, his methods always are quite perspicuous, and the motives which guided him in his investigations are evident to everyone. Clausius's exposition on the other hand always bears a trait of ceremoniousness and reserve. We are often in doubt as to whether Clausius was more concerned to tell us of something or to keep something from us. Instead of simple experiences which serve as foundations for his deductions, these deductions are built on specially assumed axioms, which have the appearance of greater reliability without really guaranteeing more than those experiences. Clausius also was addicted to creating new names and conceptions which were not always necessary. (Mach, 1986, p. 281)

It would go beyond the scope of this work to bring out the stylistic differences Mach notes and the opposition of content which separate Thomson and Clausius from one another. But, in any case, Mach's point is evidence of a symptom of significant indicative value with respect to the differences in content, on which the answer to the issue of priority depends.

It can be noted in summary that with the emergence of thermodynamics an internal pluralization in terms of content took place in the disciplinary system of modern physics, a pluralization triggered by the introduction of the concept of irreversible time. In this, priority is assumed by Thomson's essays, which highlight the finiteness and irreversibility of processes, in contrast to the works of Clausius, which are informed by the ideal of energy conservation. Yet with Thomson too this discovery takes place to a certain extent under the hand, without his having explicitly thematized their scientific or even time-theoretical dimension himself. The formulations of Coveney and Highfield, who present "Clausius as the scientific father of the arrow of time" (Coveney and Highfield, 1990, p. 150), and of Sambursky, according to whom the foundation of thermodynamics "immediately suggested the problem of the passage of time" (Sambursky, 1978, p. 41), must therefore be relativized in two respects. Not only because of the issue of priority, which Prigogine and Stengers discuss and decide in favor of Thomson, but also because an explicit treatment of the problem of time can be spoken of neither with Clausius nor with Thomson.

The position advocated by Prigogine and Stengers is to be agreed with in that, reading with hindsight, the placement of the thermodynamic research program launched by Carnot on its temporal foundations should be considered Thomson's merit. At the same time it should be highlighted more sharply than is the case with Prigogine and Stengers that Thomson lacked awareness of the significance of the step he made. Thomson did not reflect explicitly on the problem of time, nor did he become clearly aware of the confrontation between thermodynamics and dynamics—one decisive for the philosophy of science—that began to take shape on the basis of his research.

Knowledge of the plurality of scientific perspectives resulting from this confrontation as well as from the question, raised by Prigogine and Stengers at the close of the section "The Birth of Entropy" in *Order out of Chaos*, as to "how these descriptions are related" (Prigogine and Stengers, 1985, p. 122) are horizons of thought which were not yet expanded on by Thomson in his early works. The question of the relationship between dynamic and thermodynamic systems first became an issue for Thomson in the context of the debate, launched by Boltzmann in the 1870s, about the dynamic foundability of thermodynamics, to which I now turn.

## The Scientific Establishment of Thermodynamics and the Debate Concerning the Time-Theoretical Assumptions of Dynamics

The third—the most demanding and simultaneously most controversial—contribution to the foundation of thermodynamics, which at the same time signals its establishment in the disciplinary system of modern physics, was made by Ludwig Boltzmann in the last third of the nineteenth century. On the basis of the works of Carnot, Thomson and Clausius, Boltzmann turned to the problem, one not dealt with by his predecessors, of mechanically founding thermodynamics and hence to the question of the combinability of mechanics and heat theory. Boltzmann's 1872 essay "Further Studies on the Thermal Equilibrium of Gas Molecules" (Boltzmann, 1966a) was to trigger a debate opened by William Thomson (Thomson, 1966) and Josef Loschmidt (Loschmidt, 1876) in the 1870s, and taken up by Henri Poincaré (Poincaré, 1966b) and Ernst Zermelo (Zermelo, 1966a) in the 1890s. In the course of this debate the obliviousness of classical dynamics to time, which had remained concealed in the pioneering works of Thomson and Clausius, and the new problem of time posed within thermodynamics were clearly expressed.

Boltzmann's starting point was the problem, overlooked by Thomson and Clausius in their pioneering works, that he had first dealt with already in 1866 of reformulating the macroscopic irreversibility of thermodynamics at a microphysical level in the idiom of dynamics. In his essay "On the Mechanical Importance of the Second Law of Thermodynamics" Boltzmann attempted "to provide a purely analytic, perfectly general proof of the second law of thermodynamics" (Boltzmann, 1980, p. 271). This attempt distinguishes this early essay from later texts, in which Boltzmann no longer argues "purely analytically" but appeals to probabilistic arguments at the same time. The solution presented in his 1872 essay "Further Studies on the Thermal Equilibrium of Gas Molecules" (Boltzmann, 1966a) is based on a novel application to thermodynamics of the statistical mechanics that had been founded by James Clerk Maxwell (Maxwell, 1860) and which would later be brought into its canonical form by Josiah Willard Gibbs (Gibbs, 1902).

Maxwell's first step towards the mechanical interpretation of entropy consisted of rendering the state of thermodynamic equilibrium describable with the means of probability theory. Thermodynamic systems consist of a great number of molecules in rapid non-uniform motion. Because of their complexity they cannot be represented with classical methods, according to which the respective trajectory would have to be calculated for each single molecule. For this reason Maxwell took statistical averages as his starting point in calculating the velocity distribution of the molecules. By transferring the law of large numbers, applied in sociology by Quételet (Quételet, 1835; cf. John, 1968, pp. 338ff.; Stigler, 1987), to physics Maxwell succeeded in the first part of his essay in formulating the velocity distribution function named after him. A velocity distribution function determines the number of molecules with a velocity $v$—and hence kinetic energy $^1/_2mv^2$—at time $t$ (Maxwell, 1860). When the state of thermodynamic equilibrium is attained, the function assumes the form of a Gaussian curve. The equilibrium state is characterized, in the words of Maxwell in the third part of his paper, by the fact that "the collisions that incessantly modify the velocities of the molecules no longer determine any evolution in the distribution of these velocities—that is, in the mean number of particles for each velocity value" (Prigogine and Stengers, 1985, p. 240; cf. Maxwell, 1860).

Boltzmann's step beyond Maxwell consisted of not only describing thermodynamic equilibrium with statistical means, but, using the example of a rarified gas, also of formulating the evolution of the thermodynamic system towards equilibrium in probabilistic terms. To this end he developed the so called H-theorem in his 1872 essay. To begin with Boltzmann used not the letter H but the letter E for the function found by him. It was in 1895 that Boltzmann first made the change which then became canonical. Complementary to the increase in entropy, the function found by Boltzmann—which measures the change in the gas molecules' velocity distribution as a result of the collisions taking place between them—decreases uniformly in the course of time. Its minimum corresponds to that of the Maxwellian equilibrium distribution identified with thermodynamic equilibrium. From this Boltzmann concluded that the H-theorem was the microscopic expression of irreversible entropy increase which Thomson and Clausius had described at a macroscopic level in the 1850s.

Boltzmann's particular achievement consists of having linked these previously unrelated physical theories. Forms of description from classical mechanics, probability theory and thermodynamics flow into the H-theorem and form a new synthesis. Already in *From Being to Becoming* Prigogine highlights the "remarkable mixture of dynamical and probabilistic concepts" (Prigogine, 1980, p. 157) which distinguishes Boltzmann's kinetic equations. Through this novel connection, as Prigogine and Stengers summarize in *Order out of Chaos*, Boltzmann had "the key to the microscopic interpretation of entropy" (Prigogine and Stengers, 1985, p. 242) in his hands. Through Boltzmann's contribution a "principle of molecular evolution had been produced" (Prigogine and Stengers, 1985, p. 242) and in this way "a decisive step in the direction of the physics of *processes*" (Prigogine and Stengers, 1985, p. 242f.) was made.

The pioneering significance of Boltzmann's discovery, brought out by Prigogine and Stengers, was not recognized by Boltzmann's contemporaries. Directly following their previously quoted praise of Boltzmann, Prigogine and Stengers also pose the question: "Can we conclude that the problem of irreducibility has been solved, that Boltzmann's theory has reduced entropy to dynamics?" (Prigogine and Stengers, 1985, p. 243). Their conclusion agrees with that of Boltzmann's earlier critics: "No, it has not" (Prigogine and Stengers, 1985, p. 243). The limits of Boltzmann's suggested solution were clearly expressed in the debate prompted by his 1872 essay. At the same time the "clash of doctrines" (Prigogine and Stengers, 1985, pp. 233ff.) between dynamics and thermodynamics—that pre-programmed in the works of Thomson and Clausius—appeared in all severity in this debate, as did the related question of the compatibility or incompatibility of the time concepts underlying them. This second, positive, aspect is emphasized by Prigogine and Stengers in *Entre le temps et l'éternité*. By contrast, the first, negative, aspect is foregrounded in *Order out of Chaos*. The following account starts with the positive aspect, one particularly important for the theory of time, and then moves on to the negative aspect, which will provide the demarcational foil in describing the way in which Prigogine's research goes beyond that of Boltzmann.

The first reaction to Boltzmann's essay—Thomson's essay "The Kinetic Theory of the Dissipation of Energy" (Thomson, 1966), published in 1874—starts with a positive reference due to its introduction of new and important inner differentiations to physics. At the very beginning of his essay Thomson develops the distinction between classical "abstract dynamics" (Thomson, 1966, p. 177), which permits "the instantaneous reversal of the motion of every moving particle of a system" (Thomson, 1966, p. 177) and the new "physical dynamics" (Thomson, 1966, p. 177), in which "this simple and perfect reversibility fails" (Thomson, 1966, p. 177). To this end he invokes the image of "an army of Maxwell's 'intelligent demons' " (Thomson, 1966, p. 178), whose soldiers, each armed with a "molecular cricket bat" (Thomson, 1966, p. 179), were to ensure that the entropic evolution of the thermodynamic system could be undone.

In 1867 Maxwell had come up with the idea of an intelligent microscopic being which was to calculate the trajectories of a thermodynamic system's individual molecules and on this basis demonstrate the reversibility in principle of the system. Unlike Maxwell, however, Thomson commented sceptically on the consequences resulting from the underlying assumption of the universal validity of the laws of classical dynamics. Thus he emphasizes that "the real phenomena of life infinitely transcend human science" (Thomson, 1966, p. 178) and that attempted "speculation regarding the consequences of their imagined reversal is utterly unprofitable" (Thomson, 1966, p. 178). Such a reversal would, if transferred to the world and ourselves, lead to the absurd consequence—as Thomson points out with concrete human experience of time in mind—that "living creatures would grow backwards, with conscious knowledge of the future, but no memory of the past, and would become again unborn" (Thomson, 1966, p. 177f.).

The denial of irreversible time which underlies the reversibility assumption of classical dynamics is made clear by Thomson's provocative transfer of physical forms of thought to our everyday experience of time. Expressed pointedly in this way, the irreversibility underlying thermodynamics and the asymmetry of time linked with this appear to be that authority within physics which reinstates the evidence of everyday experience of time denied by classical dynamics' rigorous obliviousness to time. In this sense Prigogine and Stengers, in *Entre le temps et l'éternité*, highlight the discussion stimulated by

Boltzmann as having been "that crucial episode in the history of physics, in the course of which the mode of conceptualization of classical dynamics revealed its demands and implications" (Prigogine and Stengers, 1988, p. 25). Boltzmann's attempt to give irreversibility a dynamic meaning provoked his critics to make the implicit basic assumptions of classical dynamics explicit; that is, to reveal its constitutive underlying suppression of the irreversibility of time.

The paradoxical character of this feat of suppression becomes clear when one is reminded, following Prigogine and Stengers, that "until the end of the 19th century . . . there had been no equivalent body of ideas that would have claimed an equivalence between a plant which grows, blossoms and withers and a plant which comes back to life, becomes younger and returns to being the original seed; or between a person who matures and learns and a person who in the course of time becomes a child again, then an embryo, and finally returns to being a single cell" (Prigogine and Stengers, 1993, p. 44). Yet "dynamics," Prigogine and Stengers continue, "the physical theory which identifies itself with the very triumph of science, implied this radical negation of time. It is this which was revealed by Boltzmann's failure" (Prigogine and Stengers, 1988, p. 26).

Thomson's thought experiment, one critically directed against dynamics, of a reversal of the arrow of time is already found with Maxwell. Here, however, it is found with a thoroughly positive auspice. In a letter to his friend Lord Rayleigh, Maxwell describes an imaginary experiment in which the procession of time is reversed for all processes in our world. It is worth citing this passage in full, since it allows the absurd time-theoretical implications peculiar to the reversible worldview of classical mechanics—advocated by Maxwell—to become particularly clear. Maxwell writes

> If this world is a purely dynamical system, and if you accurately reverse the motion of every particle of it at the same instant, then all things will happen backwards to the beginning of things, the raindrops will collect themselves from the ground and fly up to the clouds, etc. etc. and men will see their friends passing from the grave to the cradle till we ourselves become the reverse of born, whatever that is. We shall then speak of the impossibility of knowing about the past except by analogies taken from the future and so on. The possibility of executing this experiment is doubtful, but I do not think it requires such a feat to upset the 2nd law of thermodynamics. (Maxwell, letter of 6/12/1870, quoted from Segrè, 1984, p. 242)

As an alternative to this scarcely implementable artifice, whose assumptions and implications he does not question, Maxwell, in a further step, introduces the demons which Thomson later picked up on:

> For if there is any truth in the dynamic theory of gases, the different molecules in a gas of uniform temperature are moving with very different velocities. Put such a gas into a vessel with two compartments and make a small hole in the A B wall about the right size to let one molecule through. Provide a lid or stopper for this hole and appoint a doorkeeper very intelligent and exceedingly quick, with microscopic eyes, but still an essentially finite being. Whenever he sees a molecule of great velocity coming against the door from A into B he is to let it through, but if the molecule happens to be going slow, he is to keep the door shut. He is also to let slow molecules pass from B to A but not fast ones. . . . In this way the temperature of B may be raised and that of A lowered without any expenditure of work, but only by the intelligent action of a mere guiding agent. . . . I do not see why even intelligence might not be dispensed with and the thing made self-acting. (Maxwell, letter of 6/12/1870, quoted from Segrè, 1984, p. 243)

As already shown, Thomson takes up the demon model in his 1874 essay. Unlike Maxwell, however, he emphasizes that this is only a theoretically possible conception. *De facto*, claims Thomson, "we can regard spontaneous disequalization as practically impossible" (Thomson, 1966, p. 182). For, he argues, with the number of molecules and the degree of their entanglement the relationship between the number of uniform molecular states and the number of non-uniform molecular states shifts increasingly in favor of the equilibrium state (Thomson, 1966, p. 182). With this Thomson anticipates the central probabilistic argument with which Boltzmann reacted to the so-called *Umkehreinwand* (reversal objection) in his 1877 reply to Loschmidt (Boltzmann, 1966b). This is the canonical concept for the objection, going back to Maxwell's demon model and explicitly raised by Loschmidt against Boltzmann, that, given the conditions of a reversal in the direction of velocities, the H-theorem must allow the reversibility of the evolution of the whole system. With reference to the case suggested by Loschmidt of a reversed evolution, that is one leading from a uniform to a non-uniform distribution of states, Boltzmann writes in his reply: "Since there are infinitely many more uniform than non-uniform distributions of states, the latter case is extraordinarily improbable and can be considered impossible for practical purposes" (Boltzmann, 1966b, p. 192).

A second objection—one which has also become canonical—to Boltzmann's attempt to explain thermodynamic irreversibility with the means of dynamics was formulated in the 1890s by Henri Poincaré. The basic problem which determined physics at the close of the nineteenth and start of the twentieth centuries is already expressed in the title of his 1893 essay: the question of the relationship between dynamics and thermodynamics or—as it was formulated in the title—between "Mechanism and Experience" (Poincaré, 1966b). Poincaré is also sceptical about the attempt to mediate offered by Boltzmann. The basic logical problem in Boltzmann's attempt, according to Poincaré, is that it is a conclusion "where one finds in effect reversibility in the premises and irreversibility in the conclusion" (Poincaré, 1966b, p. 206).

Poincaré shows that the opposite of what he attempts to derive follows from the laws of dynamics applied by Boltzmann. To this end he refers to the proof he had carried out in an earlier work (Poincaré, 1966a) "that a bounded world, governed only by the laws of mechanics, will always pass through a state very close to its initial state" (Poincaré, 1966b, p. 206). That means, however, that the evolution towards thermodynamic equilibrium cannot be an irreversible and unique process. For the initial state of this process must necessarily recur after a certain time, that is, the apparent final state of the universe must prove to be the starting point of a new evolution. Hence there cannot be a distinguished direction of time within the framework of a mechanical apprehension of the world. Finally, at the end of his essay, Poincaré praises his argument against the train of argument stemming from Maxwell as follows: "According to this theory to see heat pass from a cold body to a warm one, it will not be necessary to have the acute vision, the intelligence, and the dexterity of Maxwell's demon; it will suffice to have a little patience" (Poincaré, 1966b, p. 206).

The thermodynamic lesson from the argument stimulated by Poincaré was drawn by Ernst Zermelo in 1896. He highlights that on the grounds of Poincaré's recurrence theorem it is—against Boltzmann—precisely reversible initial states that are to be considered more probable than irreversible initial states. The distinction between theoretical possibility and practical probability, introduced by Thomson and taken up by Boltzmann, is subverted by Zermelo who commences from the assumption that "all *imaginable* mechanical initial states are physically *possible*, at least within certain limits" (Zermelo, 1966a, p. 215). From this he arrives at the result that it is impossible "on the basis of the *present* theory to carry out a mechanical derivation of the second law without specializing the initial state" (Zermelo, 1966a, p. 216f.).

Prigogine and Stengers' relationship to Boltzmann and their evaluation of the criticism levelled at him by his opponents is ambivalent. The authors highlight positively Boltzmann's intention of opening up a transition from dynamics to thermodynamics. Linked with this is Boltzmann's achievement of having contributed to the uncovering of the time-theoretical foundations of classical dynamics through the debate which he triggered. Prigogine and Stengers assess negatively the lack of implementation of his own intention and his ultimate drift into a position through which the negation of irreversible time is once again cemented within physics. This drift is a reflex reaction to Boltzmann's resigned insight into the reversible base structure of the time concept of classical dynamics revealed by his critics. Because of this insight Boltzmann gave up his original attempt to derive the arrow of time from dynamics in favor of a probabilistic argument which he understood as an approximation.

It is Boltzmann's original intention that Prigogine and Stengers adhere to in their research. The authors' position to the criticism brought forward by Boltzmann's critics results from this adherence. The relative correctness of this criticism lies in that in his original 1872 essay it was only through an intuitive linking of mechanics and probability theory that it became possible for Boltzmann to introduce time into dynamics. Thus, according to Prigogine, "unfortunately this correspondence is not 'deduced' from dynamics; it is postulated from the start" (Prigogine, 1980, p. 164). Conversely the fault in this criticism lies in its purely destructive character. The path of transforming dynamics initially pursued by Boltzmann, the aim of which was to integrate the arrow of time within the equations of dynamics, is ignored by wholesale criticism of dynamics and hence simultaneously dispensed with (Zermelo, 1966a, p. 216; Zermelo, 1966b, p. 230).

The impossibility in principle, claimed by Boltzmann's opponents, of representing thermodynamic irreversibility with the means of dynamics draws its evidential weight throughout from the supposedly invariable time-symmetrical base structure of classical dynamics. The reversal objection does this directly and explicitly by invoking the imaginary scenario of microscopically inverting the motion of the thermodynamic system's molecules, which must in principle be possible given the assumptions of dynamics. The recurrence objection does this indirectly and implicitly by taking a cyclical pattern as its basis, within which in the long term every arbitrary sequence of motion reproduces itself infinitely often. In both cases motion appears to be structurally independent of the passage of time. Time functions as an external parameter, playing a role only as a formal framework and mathematical instrument for calculating changes in the velocity of molecular elements of the system. Boltzmann's intention of proceeding against this time-theoretical reductionism at the level of dynamics itself was ironed out initially by his critics and finally by him himself.

Prigogine and Stengers, following Popper (Popper, 1992), adduce Boltzmann's second published reply, of 1897, to Zermelo's objections as obviously documenting his ultimate self-betrayal (Prigogine, 1997, p. 23f.; Prigogine and Stengers, 1988, pp. 29ff.). In addition it may be considered a preliminary stage of this failure that Boltzmann had already given up the original claim of anchoring the irreversibility of thermodynamics at the fundamental level of a dynamical account with the means of probability theory in his first reply to Zermelo. Instead of this he had fallen back on a purely probabilistic definition of entropy. Retrospectively Prigogine and Stengers highlight: "Consider once again the position with which Boltzmann was faced. He had to decide between the hope of opening up physics to temporality and his loyalty towards classical dynamics. He chose dynamics. He replaced his microscopic interpretation of the second law with a *probabilistic* interpretation connected with our lack of knowledge" (Prigogine and Stengers, 1993, p. 47).

Through this interpretation the close connection between mechanics and probability theory, which originally stood in the foreground and was defended at least with intuitive arguments, is already questioned by Boltzmann in his first reply to Zermelo. Whereas in 1872 Boltzmann had still highlighted that the conception suggested by him meant working with a "an exact theory" (Boltzmann, 1966a, p. 90), since "these probabilities can be obtained from the equations of motion alone, without having to integrate them" (Boltzmann, 1966a, p. 90f.), in the 1896 essay he emphasized "that the second law of thermodynamics is from the molecular viewpoint merely a statistical law" (Boltzmann, 1966c, p. 219) and hence "is by no means a theorem of ordinary mechanics which can be proved from the equations of motion alone" (Boltzmann, 1966c, pp. 219, 223f.).

To this Zermelo answers triumphantly in his second contribution: "Herr Boltzmann . . . wishes to change the second law into a 'mere probability theorem' which is not valid at all times. Yet he asserts that this change, whose *principal* meaning he does not misunderstand, is really unimportant" (Zermelo, 1966b, p. 230). Hence, Zermelo continues, Boltzmann—who was obviously not prepared to problematize the basic concepts of classical dynamics in favor of "a mathematical theorem, which by its nature represents only a theory which can never be directly verified" (Zermelo, 1966b, p. 230)—had spoken out against the "single principle summarizing an abundance of established *experimental* facts" (Zermelo, 1966b, p. 230) expressed in the second law. At the same time Zermelo's conclusion at the end of his essay is directed against Boltzmann's project altogether: "The great successes of the kinetic theory of gases in the explanation of *equilibrium* properties do not entail its applicability to *time-dependent processes* also, for the two are separate subjects" (Zermelo, 1966b, p. 236f.).

The opposition underlying this separation of fields, that between time-neutral dynamics and time-sensitive thermodynamics and against which

Boltzmann's original intention had been precisely directed, is then, in his second reply, explicitly dissolved by Boltzmann in favor of classical dynamics, that is, in the spirit of an option opposing the introduction of the concept of irreversible time into physics. In his rejoinder Boltzmann presents a viewpoint with "which one can understand the validity of the second law and the heat death of each individual world without invoking an unidirectional change of the entire universe from a definite initial state to a final state" (Boltzmann, 1966d, p. 242). According to this idea, which cosmologically exceeds the time-neutral horizon of classical dynamics, the world in which we live is merely a temporary and regional deviation from cosmological thermal equilibrium, in which "the two directions of time are indistinguishable" (Boltzmann, 1966d, p. 242). Prigogine and Stengers comment on this: "If at first he [Boltzmann] had wanted to interpret the irreversibility of time with the help of fundamental laws, he now traced the arrow of time back to a contingent fact" (Prigogine and Stengers, 1988, p. 30). With respect to the experience of time this fact exists in that, as Boltzmann continues, we are living beings that "find [ourselves] in such a world at a certain period of time" (Boltzmann, 1966d, p. 242)—that is, in a world far from thermodynamic equilibrium. For only in such conditions will we "define the time direction as going from less probable to more probable states (the former will be the 'past' and the latter the 'future')" (Boltzmann, 1966d, p. 242).

With this, however, Prigogine and Stengers conclude, the ground is prepared for the "subjectivist interpretation of irreversibility" (Prigogine and Stengers, 1981, p. 215), which was then cultivated by Gibbs and became the self-evident foundation for "Boltzmann's heirs" (Prigogine and Stengers, 1988, p. 31) in the first half of the twentieth century: "Boltzmann's heirs transformed what he himself experienced as a dramatic failure into a triumph. The negation of the irreversibility of time, for Boltzmann a desperate solution, became for most physicists of Einstein's generation more or less the symbol of what, for them, was the vocation of physics: getting beyond observable reality to an intelligible intemporal reality" (Prigogine and Stengers, 1988, p. 31).

Prigogine's research program is directed against this development. He understands his own research as the attempt to preserve Boltzmann's original intention from his own betrayal (Prigogine, 1997, p. 27f.; Prigogine, 1998, pp. 19ff.). Thus with regard to the projected linking of dynamics and thermodynamics Prigogine and Stengers highlight—with Boltzmann against Boltzmann: "Such a transition from the world of dynamics to the world of statistical regularities, described by probabilities, can, however, not be carried out by simply flipping over concepts which are fundamentally alien to one another. We must, so to speak, generate this transition out of dynamics" (Prigogine and Stengers, 1981, p. 247). Probability may not, as with Boltzmann and later with Gibbs and Einstein, be allowed to function as an auxiliary construction and approximation, but must itself be embedded within the foundations of dynamics.

According to Prigogine and Stengers this becomes possible when one succeeds in transforming the statistical mechanics canonized by Gibbs and Einstein so that it becomes applicable to those mechanical systems until now excluded from classical dynamics. Both Maxwell and later Poincaré have pointed out that within the world of apparently simple mechanical systems there are instabilities and non-linearities which cannot be calculated with the simple concept of the trajectory. The analysis of these systems, long neglected by physicists, but successfully carried forward in the second half of the twentieth century, has allowed a new central importance to accrue to statistical description within dynamics.

In addition, the novel anchoring of time within the foundations of physics, as supported by the non-linear extension of dynamics, has a goal which goes beyond that of Boltzmann's original undertaking. This goal results on the basis of the transformation of thermodynamics carried out by Prigogine in the 1950s and 1960s. The non-linear thermodynamics far from equilibrium developed by Prigogine has as its object systems which, unlike the equilibrium systems examined by Boltzmann, are neither isolated from their environment nor evolving towards a final entropic state. Through this new research on open, non-equilibrium-oriented systems in the field of thermodynamics the universality of "Boltzmann's Order Principle" (Prigogine and Stengers, 1985, p. 122) is called into question.

With this development the arrow of time acquires new meaning. In systems far from equilibrium time no longer appears as the entropic authority of degradation, but rather becomes a creative source of order. Within the framework of a transformed dynamics irreversibility and dissipation attain a productive dimension of meaning. The anchoring of time within the foundations of dynamics aimed for by Prigogine and Stengers should accordingly—and this is the claim of their program which goes beyond Boltzmann's original undertaking—not only found Boltzmann's equilibrium thermodynamics microscopically, but also enable a microscopic theory of thermodynamics far from equilibrium.

Prigogine's more comprehensive claim of solving "the conflict between the time concept as it is perceived within the framework of an evolutionary theory or our existential experience and the time concept of fundamental classical physics" (Prigogine, 1989, p. 49) is articulated in this second modification of Boltzmann's undertaking. The theory of dissipative structures developed within the framework of thermodynamics far from equilibrium is underlain by a time concept which transforms and differentiates the simple time arrow of classical thermodynamics in a complex manner. With the "strong temporality" (Eigen, 1984, p. 229) of dissipative structures, aspects come into play which point beyond the unidimensional time concept of classical thermodynamics towards an immanent "multitude of time" (Prigogine and Stengers, 1981, p. 287), constituting a networked framework of temporal dimensions and heterogeneous time forms. In *Order and Chaos*, Prigogine and Stengers write:

> It can be said that physics today no longer denies time and its direction. It acknowledges that the irreversible time of evolution towards equilibrium exists, the rhythmical time of structures, whose activity is fed by the currents flowing through them, the bifurcating time of the development of instabilities, indeed even the microscopic time . . . in which dynamic instability expresses itself at the microscopic level. (Prigogine and Stengers, 1981, p. 286f.)

## The Self-Organization of Time and Prigogine's Theory of Dissipative Structures

Prigogine's program of providing thermodynamics with a microscopic foundation presupposes the extension of classical thermodynamics to open systems. It proceeds from a broad, generalized concept of thermodynamics, within which the isolated systems of classical thermodynamics constitute a special case. Prigogine's works, and those of the Brussels school led by him, have for this reason found their way into the textbooks of physics under the heading "generalised thermodynamics" (Coveney and Highfield, 1990, p. 163). The recognition enjoyed by Prigogine's extension of thermodynamics within the scientific community also found expression in his being awarded the Nobel Prize, which Prigogine received in 1977 for his theory of dissipative structures.

The generalization of thermodynamics and the resultant demarcation from the narrow understanding of time and irreversibility dominating classical physics is emphasized by Prigogine and Stengers in *Entre le temps et l'éternité*:

> Every chemical reaction marks a difference between the past and the future, occurs in the direction of our future. Similarly, it is in the direction of our future, not our

> past, that heat spreads from a warmer to a colder point. However, the second law in the sense of Clausius defined this activity from a very particular point of view: such that in certain conditions it leads inevitably to its own disappearance, that is, to the state of equilibrium. (Prigogine and Stengers, 1988, p. 48f.)

Prigogine's guiding question results from this limitation of classical thermodynamics: "How can entropy generating physico-chemical activity be envisaged more generally?" (Prigogine and Stengers, 1988, p. 49).

The extension of thermodynamics to open systems, within which irreversibility plays a constructive role, took place in two stages in the course of the twentieth century. The protagonists in the first stage were Lars Onsager (Onsager, 1931) and Théophile de Donder (Donder and Rysselberghe, 1936) in the 1930s and the young Prigogine (Prigogine, 1945; 1947; Prigogine and Defay, 1944) in the 1940s. First of all in this stage an extension in perspective to open systems took place, which although prevented from attaining equilibrium states by external forces nonetheless exhibit linear behavior. Prigogine speaks in this case of "linear non-equilibrium thermodynamics" (Prigogine, 1973, p. 565).

### Linear Non-equilibrium Thermodynamics and the Creativity of the Arrow of Time

The object domain of linear non-equilibrium thermodynamics comprises systems which, following destabilization through external energy and material currents lying below a critical threshold, cannot reach the entropy maximum, yet at the same time (due to the slight degree of destabilization) retain the evolutive orientation towards a final state. In these cases one speaks of systems close to equilibrium which, since they cannot themselves attain equilibrium, strive for a state resembling equilibrium: a stationary state. This state can be calculated by means of a potential function (like the equilibrium state in isolated, or closed, systems). The evolution of the system is hence determined by the orientation towards a given minimum or maximum and is therefore precisely predictable.

Because of the calculability and stability of these systems, the role played by time in dynamic equilibrium systems does not differ fundamentally from the role assumed by time in classical thermodynamic systems. This, in any case, is the assessment given by Prigogine and Stengers in *Order out of Chaos*. Just as in the equilibrium state, time is also at work in the stationary state as an irreversible factor of a continual increase in degradation. The theorem of minimal entropy production, formulated by Prigogine in 1945 to describe stationary dynamic equilibrium (Prigogine, 1945), expresses, according to

Prigogine and Stengers in *Order out of Chaos*, "a kind of 'inertia' " (Prigogine and Stengers, 1985, p. 139). If the boundary conditions prevent the system from attaining equilibrium, then, Prigogine and Stengers continue, "it does the next best thing; it goes into a state of minimum entropy production—that is, to a state as close to equilibrium as 'possible' " (Prigogine and Stengers, 1985, p. 139). That means that in the course of time, just as with the evolution towards equilibrium, the system evolving towards a stationary state "forgets" the initial conditions from which its evolution started. Below the critical threshold, varying initial conditions and contingent fluctuations do not affect the system, which is predetermined by a uniform and hence calculable orientation towards the stationary state. Prigogine and Stengers' resumé in *Order out of Chaos* thus reads: "We see that in the linear range the situation remains basically the same as that of equilibrium" (Prigogine and Stengers, 1985, p. 139).

In contrast to this, a modified assessment of the significance of linear non-equilibrium thermodynamics for the question of time is provided by Prigogine and Stengers in *Entre le temps et l'éternité*. Here the stationary state is no longer interpreted in terms of the equilibrium model. Conversely the equilibrium state appears instead "as a particular example of a stationary state" (Prigogine and Stengers, 1988, p. 49). The stationary state is characterized by the fact that the external supply of energy and matter, leading to a negative entropy flow, is permanently compensated for by internal entropy production. This means that in the stationary state the system remains constantly active: "entropy producing activity continues to occur, maintained by continuous exchange with the environment" (Prigogine and Stengers, 1988, p. 49). By contrast, the specific feature of the equilibrium state is that exchange processes with the environment either do not occur (isolated system) or effect no change in entropy (closed system). For this reason the entropy production in the equilibrium state is equal to zero. In contrast to this, starting from a stationary state opens up a more general description insofar as it allows "entropy producing activity itself to be characterized, and not from the sole perspective of its disappearance at equilibrium" (Prigogine and Stengers, 1988, p. 49).

The creative power which lies in entropy producing activity already appears in systems close to equilibrium. This is shown by Prigogine and Stengers in *Entre le temps et l'éternité* using the example of conduction in a system consisting of two vessels. Each of the two vessels is filled with a mixture of two gases (e.g. hydrogen and nitrogen). The system is initially in equilibrium. It is then destabilized with one container being constantly heated and the other constantly cooled off. Prigogine and Stengers describe the result of this simple experiment as follows:

> The experiment shows that, coupled with the process of conduction, a process of *separation of the two gases* occurs. When the system has reached its stationary state, so that for a given heat flow the temperature difference no longer changes with time there will, we say, be more hydrogen in the warm container and more nitrogen in the cool container, with the difference in concentration being proportional to the difference in temperature. (Prigogine and Stengers, 1988, p. 49f.)

This creative aspect, which already plays a role at the level of linear non-equilibrium thermodynamics, is not considered by Prigogine and Stengers in *Order out of Chaos*. The reason for this is that in *Order out of Chaos* the system's tendency to minimize the difference in heat is in the foreground. A feat of ordering, possibly linked with this, which precipitates itself in other parameters (concentration distribution) is not considered in *Order out of Chaos*. Because of the coupling of heat flow with diffusion in the example quoted it is precisely in this second respect, however, that a difference is produced. Parallel to the minimization of the difference in heat a "process of anti-diffusion" (Prigogine and Stengers, 1988, p. 50) takes place which represents "a *negative* contribution to entropy production" (Prigogine and Stengers, 1988, p. 50). A "negentropic" orientation arises within the system which can no longer be grasped with the means of classical thermodynamics, from whose perspective entropy production is always either positive or equal to zero.

According to this extended view, the theory of stationary states can already contribute to freeing thermodynamics from the classical idea "that entropy producing activity is synonymous with degradation" (Prigogine and Stengers, 1988, p. 50). Unlike in *Order out of Chaos*, in which they highlight that for irreversible processes "in the linear range the situation remains basically the same as at equilibrium" (Prigogine and Stengers, 1985, p. 139), Prigogine and Stengers hence come to the conclusion in *Entre le temps et l'éternité*: "The study of stationary states suffices to dissociate the second law from the idea of evolution towards 'disorder,' inertia, and uniformity" (Prigogine and Stengers, 1988, p. 49). With this the first step towards overcoming the limits of classical thermodynamics, that made by Onsager, Donder and the young Prigogine, is revalued in time theoretical terms since in this "the creative and destructive duality of irreversible processes" (Prigogine and Stengers, 1988, p. 51) is expressed.

However, going beyond linear non-equilibrium thermodynamics, in *Entre le temps et l'éternité* the particular significance assumed by non-linear thermodynamics is emphasized at the same time. The emergence of order in systems evolving linearly towards a stationary point is characterized by the fact that the ordered structures emerging can be linearly traced back to the external forces to which the system is subjected. In the example of conduction, the question as to the price to be paid for the creation of order is easily answered: "The chemical separation of the two gases is not a sorting carried out once and for all, but the result of a process which is paid for through the permanent creation of 'disorder,' with the levelling of the temperature difference which maintains the heat flow" (Prigogine and Stengers, 1988, p. 51). The generation of order in systems close to equilibrium is directly compatible with the classical reading of the second law. There is no excess of order, rather the negentropy being generated can be directly traced back to a corresponding use of energy. Hence it holds that:

> Close to equilibrium this constraint, imposed by us, suffices to determine the activity of the system: the stationary state in fact corresponds to the minimal activity which is compatible with the constraint keeping the system from equilibrium. . . . One can therefore say with a process like that of thermodiffusion that the thermal difference imposed on the system "explains" its activity. This will not be the same beyond the instability threshold. (Prigogine and Stengers, 1988, p. 59)

The task of non-linear dynamics far from equilibrium is to make understandable the generation of order in systems in which the ordering achieved by the system unfurls a dynamics of its own that can no longer be reduced in direct proportion to the external forces to which the system is subjected. The actual achievement of the second step in the transformation of thermodynamics lies in this. Beyond its creative dimension of meaning irreversible time thus reaches a new kind of independence from external forces that is linked with a "sensitivity" for internal changes as well as for differences, aspects and factors of the system environment that remain irrelevant when close to equilibrium.

### Non-linear Non-equilibrium Thermodynamics and the Temporality of Dissipative Structures

In the 1950s and 1960s Prigogine began to extend his research to the unstable behavior exhibited by systems far from equilibrium. In doing this his hope initially was to establish the applicability of the theorem of minimal entropy production, valid for stationary states, for situations far from equilibrium too. It turned out, however, that "in systems far from equilibrium the thermodynamic behavior could be quite different—in fact, even directly *opposite* that predicted by the theorem of minimal entropy production" (Prigogine, 1980, p. 88). This demarcation accentuates once again the side of stationary state theory amounting to degradation and the levelling of differences that was foregrounded in *Order and Chaos*. Against this, from the perspective of *Entre le temps et l'éternité*, the statement that behavior far from equilibrium is "quite unlike" behavior close to equilibrium must be made more precise. For close to equilibrium dissipation and irreversibility can lead to the generation of order—as the example of a diffusion process coupled with conduction has shown.

The specific difference between systems far from equilibrium and systems directed towards stationary states consists of the fact that close to equilibrium the generation of order can be directly traced back to external forces, whereas far from equilibrium this is no longer the case. However, self-organization—as Prigogine and Stengers define it in *Entre le temps et l'éternité*—is first found when the behavior of the system is no longer fully determined by the boundary conditions and we can attribute it a "certain autonomy" (Prigogine and Stengers, 1988, p. 59). According to this, the specific difference which distinguishes the generation of order close to equilibrium from the generation of order far from equilibrium is the difference between an externally determined organization and internal self-organization. Prigogine comments: "Below this threshold, which represents the end of the domination by an attractor linked with an equilibrium state, dissipation serves as a source of order and new creation. This occurs not only through the superposition of two processes, with one abolishing a differentiation while the other creates a new one"—as is the case in the example of thermodiffusion, "but also through a really collective process of self-organization in which periodic behaviour—spatial symmetry breaks etc. appear. This is the domain of 'dissipative structures' " (Prigogine and Stengers, 1990, p. 12).

In addition to this comes the fact "that the laws of fluctuation close to equilibrium are *universally valid*, whereas at greater distance from equilibrium they become specific with a non-linear kinetics according to the type of non-linearity concerned" (Prigogine and Stengers, 1981, p. 180). Whereas close to equilibrium a linear relationship exists between temperature and the concentration difference, in systems driven so far from equilibrium that they have lost the direction towards a stationary point "sudden, spectacular phenomena" occur "such as the appearance of new, qualitatively different, regimes of behaviour" (Prigogine and Stengers, 1988, p. 52).

This has consequences for the concept of time: "Stationary states belonging to the thermodynamic branch can be described fully through their composition and boundary conditions. As the expression of a deterministic evolution they retain no memory of the system's past. By contrast, a dissipative structure is receptive to a real history linking the random play of fluctuations and the determinism of the laws of average" (Prigogine and Stengers, 1990, p. 13). The theory of dissipative structures brings into play time concepts that dissolve the uniform, goal-directed time concept of classical thermodynamics into a plurality of narrative temporality structures and system internal times. Concepts like "suddenness," "instant," "event" and "history" become central metaphors for describing the thermodynamic concept of time far from equilibrium. This pluralizing and historizing extension of the irreversible time concept can be made clear using several spectacular examples from the realms of hydrodynamics, chemistry and physics.

The non-linear dimension of the evolution of open systems, which takes shape beyond the critical threshold, had already been commented on frequently in scientific history before being examined in detail by Prigogine in the 1960s. Unstable chemical or hydrodynamic systems were for a long time considered special cases and curiosities, or were disqualified as being the elaborations of charlatans, so that research into them wasn't seriously undertaken and the scientific mainstream drew no conclusions from their existence.

A simple example of such a system stems from hydrodynamics and is linked with the phenomenon of convection, which had already been studied from the eighteenth century onwards. The first scientific descriptions of the phenomenon are found in the works of Rumford, dating from 1798, on heat transport in an apple pie (Chandrasekhar, 1961, p. 9; Velarde and Normand, 1980, p. 119). Convection currents are a structured flow motion that can arise in gases or fluids to balance out internal density or temperature differences. At the beginning of the twentieth century Henri Bénard had already attempted to research systematically these currents in his essay "Circular Vortices in a Liquid Layer" (Bénard, 1900). Against the background of Prigogine's works they have acquired new, central importance. In addition they have in the meantime been reconstructed in detail using computer simulations (Mareschal and Kestemont, 1987).

Bénard convection consists of hexagonally ordered patterns arising in a thin layer of liquid, heated from below, when the supply of heat oversteps a certain critical value. From the perspective of the equilibrium concept the sudden generation of this coordinated behavior of numerous molecules over a large range is completely unexpected. Instead it would be plausible from the equilibrium perspective for the molecular motion through which heat is transported to become ever more lawless with increasing heat supply. This idea, however, holds only for systems below the critical threshold. Above the threshold the system begins to transport heat not only through intermolecular collisions (conduction), but rather through convection, that is, through the collectively ordered motion of whole groups of molecules within which the molecules move with uniform velocity. In this case the system's stationary state is characterized by heat transfer through convection. The system's temperature, density and pressure are not uniform as in the homogeneous equilibrium state, but vary linearly from bottom to top between warm and cold regions.

This self-organization of the system is accompanied by a correlation of wide reach. Whereas the correlations characteristic for equilibrium states extend over around one Angstrom ($10^{-10}$m), the length of a Bénard cell is of the order of millimeters ($10^{-3}$m). For such a cell it holds that "molecules drawn into a vortex can no longer be defined as units independent of one another" (Prigogine and Stengers, 1988, p. 53). This contradicts Boltzmann's order principle and hence the second law in its classical version. Whereas a multitude of microstates—for example, divergent molecular velocities—lead macroscopically to an increase in disorder, there are only few microstates—for example, the correlated motion of molecules—which give rise to a macroscopic order. According to the law of large numbers the probability of the appearance of Bénard convection is close to zero. All the same, it appears regularly once the critical threshold value is reached.

The appearance of convection is precisely predictable and deterministic. The specific structure of the convection, however, is not predictable. Nicolis and Prigogine comment on this:

> On the one side the experiment is fully reproducible. . . . On the other side, however, the matter is structured in cells that . . . rotate alternately clockwise or anticlockwise. As soon as a certain sense of rotation has established itself, this continues to remain unchanged. Nonetheless, just above the critical threshold $\Delta T_k$ two qualitatively different situations can appear, quite independently of how refined the control of the experiment might be. (Nicolis and Prigogine, 1987, p. 27)

Because of this bifurcational character of the critical threshold, which incorporates two possibilities, the prediction of the critical point can be called the prediction of the unpredictable. Nicolis and Prigogine add to this: "Actually one has even to speak of *infinitely many* possibilities, since in a very large system the structure can be shifted horizontally by an arbitrary amount" (Nicolis and Prigogine, 1987, p. 27). And, summing up, Prigogine and Stengers write: "In the case of the Bénard instability it is a fluctuation, a microscopic convection current, which would have been doomed to regression by the application of Boltzmann's order principle, but which on the contrary is amplified until it invades the whole system" (Prigogine and Stengers, 1985, p. 143).

In *Researching the Complex* Nicolis and Prigogine point out the change in structure of space and time that can be demonstrated already in this simple case of an unstable system. Whereas a tiny observer observing the states of single volume elements of the liquid layer will be able to ascertain no spatial or temporal difference in the equilibrium state, this changes the moment the liquid begins to organize itself. From the homogeneous space-time structure of the equilibrium system, which makes it impossible for an imaginary observer to develop "a concept of time from within" (Nicolis and Prigogine, 1987, p. 23) or "a perception of space" (Nicolis and Prigogine, 1987, p. 22), there arises far from equilibrium a qualitatively structured system spatiality and temporality. The observer "can now tell where he is by orienting himself according to the sense of rotation of the cell in which he currently finds himself. In addition he can obtain a pretty effective idea of space by counting the number of cells he has passed through when walking around" (Nicolis and Prigogine, 1987, p. 25). And time now also acquires a qualitative dimension. In that a contingent microscopic fluctuation—that is, in that a temporary microevent can inscribe itself in the macroscopic evolution of the system—the system acquires "a historic dimension" (Nicolis and Prigogine, 1987, p. 28). The memory of a past event at the moment the convection came about—consisting of the contingent selection of a particular direction of rotation (there are clockwise and anticlockwise Bénard cells)—is preserved by the macrostructure of the system and determines its further evolution.

Whereas in the equilibrium system each fluctuation is levelled by, and reduced to, the domination of averages, "the fact that an event can 'attain meaning,' that is, can cease being a mere rustle in the senseless turmoil of microscopic activity, introduces that narrative element to physics" (Prigogine and Stengers, 1988, p. 61) which leads to a pluralization of time into a multitude of self-organizing system internal times. The "sensitivity" (Prigogine and Stengers, 1988, p. 60) expressed in this, which characterizes systems far from equilibrium, refers to both internal fluctuations and external parameters that play no role in equilibrium, but which can gain decisive influence over the system's identity far from equilibrium.

In the case of Bénard convection it is the system's novel sensitivity to gravitation which leads to the possibility of a build up of internal fluctuations. Below a critical threshold the stabilizing effects of the liquid's viscosity and heat transfer by conduction oppose the destabilizing effect of the heat supply. But as soon as the critical threshold is crossed, the destabilization of the system means that the liquid elements in the hotter, lower region have a lower density than those in the colder, upper region. As a result the system's temperature gradient finds itself in "contradiction" (Prigogine and Stengers, 1988, p. 59) with gravity, which causes an inverted density distribution. The way in which this situation can lead to instability and finally to the system's self-organization is described by Nicolis and Prigogine as follows: "To do this just consider a small element of liquid close to the lower plate and imagine that it is shifted upwards slightly by some disturbance. Since it now finds itself in a colder and hence more dense environment, it experiences an Archimedean uplift which amplifies the upwards movement still further" (Nicolis and Prigogine, 1987, p. 24). The same happens in the opposite direction. As the system becomes sensitive to the effects of gravity through thermal destabilization, it can happen that the interference of Brownian motion—irrelevant in equilibrium—builds up and imposes a new, ordered regime on the system.

A close connection between sensitivity and identity of systems is demonstrated in this. "It is the system's activity that 'gives meaning' to gravity, which integrates this in a specific manner its in its own functional mode and hence gravity provides the system with the possibility of developing new structures, new differentiations" (Prigogine and Stengers, 1988, p. 60). The system's identity cannot, as in the case of equilibrium or states close to equilibrium, be defined through a reduction of the boundary conditions to a series of manipulable factors. The definition of the system's identity is far more dependent on its inner activity, which, for its part, incorporates both external and internal factors in an unpredictable manner. "The notion of 'sensitivity' links what physicists had previously been used to separating: the definition of the system and its activity. . . . It is the system's intrinsic activity that determines how we are to describe its relationship to the environment, which therefore engenders the kind of explanation which will be pertinent in understanding its possible histories" (Prigogine and Stengers, 1988, p. 60).

Hydrodynamic self-organization phenomena, however, disappear as soon as the external forces cease or are reduced. The system switches back to the equilibrium state and the ordered structures dissolve. This also applies fundamentally for the chemical processes on which Prigogine's self-organization theory is based. But, in contrast to the hydrodynamic Bénard instability, given certain conditions in chemical systems "new material structures [can arise], which to a certain extent constitute traces and witnesses of the conditions of their own formation" (Prigogine and Stengers, 1988, p. 85). At the same time whereas Bénard convection is attributed not to chemistry, but to physics since "the chemical composition of the substances in the layer does not change during all processes" (Nicolis and Prigogine, 1987, p. 29), with chemical clocks no "simple mechanical origin" (Prigogine and Stengers, 1985, p. 144) can be stated that brings about the instability. All the same, the time-theoretical implications of the theory of dissipative structures can be better clarified using examples from the domain of chemical processes.

The two historically most important examples of self-organization processes in the field of chemistry are the Belousov-Zhabotinsky reaction in the experimental domain, and the so-called Brusselator in the domain of theoretical models. The Belousov-Zhabotinsky reaction was discovered by chance by Boris Pawlowitsch Belousov in the 1950s while he was attempting to develop a simple laboratory version of the citric acid cycle. When, as part of his attempts, he dissolved citric acid and sulphuric acid together with potassium bromate and a cerium salt in water, a regularly recurring change in color between yellow and transparent set in. Anatol Markovich Zhabotinsky examined this reaction further in the 1960s and brought it into the form usual today by replacing the cerium compound with an iron compound. From this resulted the—often demonstrated—spectacular periodic switch in color between blue and red.

The Belousov-Zhabotinsky reaction was the first spatially oscillating and temporally periodic reaction—a so-called chemical clock—to be scientifically investigated and finally also acknowledged by the scientific community. Belousov's first account of his discovery was in a manuscript from 1951 (Belousov, 1987b). However, the text was turned down by the editors of the scientific journal to which Belousov had submitted it. Only in 1959 did a two-page article appear in the documentation of a symposium on radiotherapy (Belousov, 1987a). Since the reaction described by Belousov was considered to be incompatible with the second law, it wasn't until the year 1968, on the occasion of a Prague conference, that Belousov's discovery—which had been examined more closely by Zhabotinsky in the 1960s (Zhabotinsky, 1987)—received international acknowledgement and was related to phenomena known from biology.[3]

Just how unexpected the appearance of chemical clocks must seem against the background of the equilibrium conception is emphasized by Prigogine and Stengers in their description of the Belousov-Zhabotinsky reaction:

> Suppose we have two kind of molecules, "red" and "blue." Because of the chaotic motion of the molecules, we would expect that at a given moment we would have more red molecules, say, in the left part of the vessel. Then a bit later more blue

> molecules would appear, and so on. The vessel would appear to us as "violet," with occasional irregular flashes of red or blue. However, this is *not* what happens with a chemical clock; here the system is all blue, then it abruptly changes its color to red, then again to blue. Because all these changes occur at *regular* time intervals, we have a coherent process. (Prigogine and Stengers, 1985, p. 147f.)

Until and during the 1960s, chemists, following the second law, had assumed that chemical reactions in reaction glasses fundamentally exhibit no reproducible periodic behavior, but move constantly towards a homogeneous and time-independent final state. All deviations from this rule were traced back to deficiencies in the performance of the experiment or even to consciously deceptive maneuvers, and were hence not made the object of scientific research. This was the case, say, for the pioneering works already published in the 1920s by William Bray about a chemical oscillation he had discovered (Bray, 1921). Further preliminary steps toward the research of chemical structure formation are found in the works by Friedlieb Ferdinand Runge and Raphael Eduard Liesegang (Kuhnert and Niedersen, 1987).

In fact it can be said that "in the range of validity of the uniform stationary state the system . . . simply ignores time. But as soon as it finds itself in the range of periodic states of oscillation it quite suddenly 'discovers' time in the periodic motion's phase and in the fact that the maxima of the different concentrations follow one another according to a given temporal scheme. We could call this *temporal symmetry breaking*" (Nicolis and Prigogine, 1987, p. 37). The sequence of states which the system cyclically reproduces is in fact in no way reversible. Rather a certain sequence of processes is distinguished by the chemical clock. There is an asymmetry between future and past: "Chemical clocks always go in one direction. Assuming the chemicals A, B, C and D were concerned and that we saw the sequence A → B → C → D → A. The remarkable fact is that the 'opposite' sequence A → D → C → B → A does not occur. This difference obviously goes beyond the mere ascertainment of an approximation to the probable state" (Prigogine and Stengers, 1981, p. 246; for details see Andronov and Chaikin, 1949).

In addition, the system develops a temporal rhythm of its own which makes it possible to use it as a clock. This is not the case with Bénard convection, in which a single contingent situation inscribes itself in the further evolution of the system. This evolution, however, is not characterized by a temporal pattern of its own. In the case of the Belousov-Zhabotinsky reaction things are different. Here the system reproduces its own state of instability according to a regular periodicity so as to oscillate to and fro between two ordered states:

> Far from equilibrium, therefore, an unexpected relation exists between chemical kinetics and the *space-time structure* of reacting systems. . . . Therefore we may say that chemical instabilities involve *long-range order* through which the system acts *as a whole*. This global behavior greatly modifies the very meaning of space and time. Much of geometry and physics is based on a simple concept of space and time, generally associated with Euclid and Galileo. In this view, time is homogeneous. Time translations may have no effect on physical events. Similarly, space is homogeneous and isotropic; again translations and rotations cannot alter the description of the physical world. It is quite remarkable that this simple conception of space and time may be broken by the occurrence of dissipative structures. Once a dissipative structure is formed, the homogeneity of time, as well as space, may be destroyed. We come much nearer to Aristotle's "biological" view of space-time. (Prigogine, 1980, p. 103f.)

Furthermore, the chemical reactions differ from the hydrodynamic currents in that they do not—as do the latter—immediately become turbulent at correspondingly great distances from equilibrium. For chemical processes, far from equilibrium conditions are a necessary, but not yet a sufficient, condition for microscopic instabilities. Chemical systems far from equilibrium first become unstable when the further condition that an autocatalytic process is built into them is satisfied. One speaks of "autocatalysis" in chemistry when the product of a chemical reaction participates in its own synthesis (Prigogine and Stengers, 1985, p. 143f.). The concept goes back to Wilhelm Ostwald (Ostwald, 1890). The scientific examination of autocatalytic feedback structures was developed especially within the framework of cybernetics and general system theory (Jantsch, 1980, p. 5f.). The relationships through which chemical clocks function and which the chemical mechanism seeks to explain were revealed by the so-called Brusselator model—introduced by Prigogine and Lefever at the end of the 1960s (Prigogine and Lefever, 1968) and further developed by Prigogine in the 1970s—and later reconstructed in detail by Epstein et al. using computer simulations (Epstein, Kustin et al., 1989).

The Brusselator is a model which may be used to examine in simple conditions the multitude of possible effects that can result in autocatalytic chemical reactions far from equilibrium. The model works with two chemical substances A and B which pass through four intermediate steps, in which the substances X and Y participate, and result in the products D and E. A schematic account of the four intermediate stages is found in *Order out of Chaos* (Prigogine and Stengers, 1985, p. 146f.). In the first step a molecule X is produced from molecule A. In the second step X reacts with a molecule B, generating Y and the product D. In the third step the combination of two

X molecules with a Y molecule produces three X molecules. This is the autocatalytic stage of the reaction that is responsible for the instability: the X molecule catalyzes its own formation and is hence sensitive to feedback. In the fourth step the direct transformation of X into the product of the reaction takes place. The system is kept far from equilibrium permanently through the continuous provision of the substances A and B and continuous removal of the products D and E, with the reaction glass always being well agitated at the same time.

Using the Brusselator changes in the concentrations of X and Y are examined as they depend on the throughput imposed on the system. At low throughput values of A and B the system finds itself close to equilibrium. Crossing the first critical threshold leads to the periodic behavior, familiar from the Belousov-Zhabotinsky reaction, of the so-called limit cycle. The concentrations of X and Y move in regular orbits about the now unstable stationary state (Coveney and Highfield, 1990, p. 192; Prigogine and Stengers, 1985, p. 147). The limit cycle itself represents a new stable state from which the system cannot extract itself through fluctuations. The trajectories of all possible initial states for the given boundary conditions are drawn towards the limit cycle as an attractor and lead to the chemical clock's periodic behavior. To this extent it can be said that "a system characterized by a limit cycle remains a predictable system that can be described in a simple manner" (Prigogine and Stengers, 1988, p. 72).

The limit cycle differs from simple attractors, such as the equilibrium state for example, only in that it is an attractor which takes the form not of a point, but of a line. The linear attractor is described with the help of the so-called Hopfian bifurcation (Hopf, 1942). This is a simple branching in which the system is unstable only at the bifurcation point itself, but then immediately switches to a new, stable—the periodic—state. Prigogine and Stengers note: "Far from exhausting the new solutions that may appear, this primary bifurcation introduces only a single characteristic time (the period of the limit cycle) or a single characteristic length. To generate the complex spatial temporal activity observed in chemical or biological systems, we have to follow the bifurcation diagram further" (Prigogine and Stengers, 1985, p. 167).

The system's complex transitional structure, that is, the linking of a plurality of different stable and unstable states in time, is represented using the cascade model—so-called Turing bifurcation—developed by Alan Turing in 1952 and adopted by Prigogine and Lefever. Analysis of the Brusselator falls in the range of these complex structures when diffusion within the system is taken into account. At the experimental level of the Belousov-Zhabotinsky reaction this means that the reaction is now carried out without regular mixing (that is without agitation). In this case regular spatio-temporal patterns occur beyond the critical threshold whose evolution can no longer be as easily predicted as that of the chemical clock (cf. the illustrations in Nicolis and Prigogine, 1987, p. 39f.). Correspondingly in the Brusselator model "the diffusion of the chemical throughout the system induces, in the far from equilibrium region, the possibility of new types of instability, including the amplification of fluctuations breaking the initial spatial symmetry. Oscillations in time, chemical clocks, thus cease to be the only kind of dissipative structures available to the system" (Prigogine and Stengers, 1985, p. 148). The ordered spatial patterns can of course link with the temporal oscillations of the chemical clock:

> For instance the limit-cycle resulting from a Hopf instability can turn in space as well as time. One familiar object which changes in space and time is a wave—think of one breaking on a pebble beach. And indeed, in a chemical clock where Hopf instability rules, we should expect to see ripples of red and blue passing through the reactor rather than the entire solution in the reactor simultaneously changing colour to red or blue instantaneously. (Coveney and Highfield, 1990, p. 196)

At the level of systems of complex bifurcations that go beyond "primary bifurcation" (Prigogine and Stengers, 1985, p. 167)—that is, beyond simple Hopf bifurcation—which result when diffusion is also considered and the system's throughput further increased, ever more complex temporal and spatial structures develop. Particularly interesting in our context is the fact, already highlighted in the section on Bénard convection, that multiply bifurcated systems far from equilibrium unfurl a historical dimension. Prigogine and Stengers comment:

> Once the system is moved further and further from equilibrium through interaction with the environment, the system passes through zones of instability towards certain fluctuations (bifurcations) and its evolutionary path can assume a quite eventful "historical" character. In the bifurcation ranges, fluctuations decide which working regime the system will subsequently find itself in. Hence, alongside its composition and the boundary conditions, the determinant quantities of a non-equilibrium system also include its own history. (Prigogine and Stengers, 1990, p. 14)

The close connection between bifurcation and narrative temporal structure intensifies with the degree of bifurcation tree's complexity: "The 'historical' path along which the system evolves as the control parameter grows is characterized by a succession of stable regions, where deterministic laws dominate, and of instable ones, near the bifurcation points, where the system can 'choose' between or among more than one possible future" (Prigogine and Stengers, 1985, p. 169f.). However, as can be shown using the Brusselator, this "mixture of necessity and chance" (Prigogine and Stengers, 1985, p. 170), characteristic of historical developments, is part of an intermediate region. It arises at a certain distance from equilibrium and breaks down again as soon as the distance from equilibrium becomes too great. Prigogine and Stengers write: "It has been observed for a certain bifurcation diagram that one can even attain behaviour with which, having passed through a series of subsequent bifurcations, the system becomes chaotic" (Prigogine and Stengers, 1981, p. 170). From this the authors conclude: "The distance from equilibrium must be large enough, but may not be too great, if one is to avoid destruction of the delicate structure necessary for the maintenance of the normal functions of life" (Prigogine and Stengers, 1981, p. 171). The cyclic temporal structures of life, and the historical dimensions and temporally interlocking structures of individual and collective chemical and biological systems are an extremely sensitive and precarious phenomenon representing "a kind of sandwich level between the thermal chaos of equilibrium and the turbulent chaos of non-equilibrium" (Prigogine and Stengers, 1981, p. 170).

## The Objective Temporalization of Time in Physics and the Concept of Irreversible Time

The preceding account of scientific history has shown how the introduction of the concept of the irreversible arrow of time, which took place within the framework of the emergence and scientific establishment of thermodynamics in the disciplinary system of physics, lead to an internal pluralization of physical discourse. From the attempt to operationalize complex thermodynamic systems with the means of physical formalization, the necessity resulted within physics to supplement the reversible time concept underlying classical mechanics with modified time concepts that imply in various ways the irreversibility of a time directed from the past into the future. The transition thus expressed from a Newtonian worldview, one oriented towards universality and uniformity, to a science operating with plural models of the construction of physical objects had already been highlighted by Comte in the middle of the nineteenth century from the perspective of the philosophy of science. Within science itself the transition from a uniform conception of physics to a pluralistic one working with heterogeneous time concepts became a theme of the discussion, prompted by Boltzmann, about the possibility or impossibility of a mechanical theory of thermodynamics.

Against this background Boltzmann's original project, in whose succession Prigogine's research is situated, proved to be an undertaking which aimed to anchor firmly the irreversible arrow of time beyond the macrophysical realm of thermodynamics—in which it is indispensable—within physics at all levels. This expresses Prigogine's claim to be developing a uniform physical theory of the complex encompassing all phenomena, one freed from all the internal conflicts arising from the confrontation of varying time concepts. At the same time the preceding account of scientific history has made clear that Boltzmann's original project, to which Prigogine appeals, must be contrasted with his later climbdown position. This means recognizing the fact that the irreversibility of time cannot be anchored in mechanics, but is the effect of a statistical description forced by the system's complexity, which has the character of an approximation and could, in principle, be theoretically reduced to reversible trajectories in the microscopic description.

The two interpretations of irreversibility—the microscopic theory of irreversibility, advocated by Prigogine in recourse to Boltzmann's original project, and the macroscopic theory, which results from Boltzmann's climbdown position and has been developed in the philosophy of science by Reichenbach (Reichenbach, 1971, 1957) and Grünbaum (Grünbaum, 1973) among others—stand in unreconciled opposition to one another in current discussion (Kroes, 1982, esp. pp. 117–123; Kroes, 1985, esp. pp. 147–176). Boltzmann's later position, that emphasizing the approximative character of macrophysical descriptions, has gained plausibility through the transformation of the deterministic H-theorem into a time-symmetrical statistical formulation—the so-called generalized H-theorem (Davies, 1974, pp. 29–79, esp. pp. 49–51, 60–62). On the other hand, through both the described extension of thermodynamics to systems far from equilibrium and the transformation experienced by dynamics within the framework of chaos theory and the introduction of fractal mathematics,[4] the borders between microphysics and macrophysics have started to shift in a way underpinning the prospect, provided by Prigogine and Stengers in the *Paradox of Time*, of "a consistent formulation, including irreversibility and probability, of the fundamental laws of physics" (Prigogine and Stengers, 1993, p. 33; cf. also Prigogine, 1998).

The question as to which of the two opposing physical irreversibility theories will hold up and assert itself in the future must be left open here. It should be emphasized, however, that through their universalist self-interpretation, Prigogine and Stengers provide the microphysical theory of irreversibility they are striving for with philosophical implications that cannot be covered by the facts and which fall behind the pluralistic perspective resulting from their scientific works. This can also be seen in those passages of their reconstructions of scientific history in which the authors smoothly switch from the reconstruction of facts to the universalistic interpretation thereof.

An example of this switch is provided by the accusation, directed by Prigogine and Stengers against classical physics, that its operational procedures amount to a denial or obliviousness to time. With Prigogine and Stengers the seemingly purely descriptive talk of the "elimination of time" (Prigogine, 1998, p. 17) or the "denial of the arrow of time" (Prigogine and Stengers, 1993, pp. 9, 42ff.) implies a realistic correspondence theory of truth. According to this theory physics falls short of its supposed task of depicting external reality by abstracting from the irreversibility of time (which Prigogine and Stengers locate in a reality encompassing humans and nature).

This accusation should be contrasted with the constructivist self-understanding possessed by modern science since Descartes (Heidegger, 1977b). The idea, highlighted by Prigogine and Stengers, that the alleged obliviousness to time was noticed only at the end of the nineteenth century can be straightforwardly explained against the background of this self-understanding. A science understanding itself constructively and not realistically will, as a rule, question its underlying base assumptions and basic paradigms neither because of their incompatibility with a reality supposed to be theory-independent, nor through a comparison with everyday experience postulated as being theory-independent (Bachelard, 1975). Scientific scrutiny of its own assumptions begins far more when a theory appears that replaces the basic assumptions of the old paradigm with other assumptions. Thus in terms of the history of science it is in no way surprising that it was the outbreak of the dispute between thermodynamics and dynamics which first led to the explicit thematization of the concept of reversible time underlying classical physics. Previously there had quite simply been no serious competitors to question precisely this assumption.

But a second implication is linked with the talk of the denial of time by classical physics, one resulting from Prigogine and Stengers' tendency to switch directly from the descriptive pluralism of their reconstructions of scientific history to the speculative universalism of their philosophical self-interpretation. This implication lies in their suggestion that "an original concept" (Prigogine and Stengers, 1981, p. 268)—one leading science back to its actual essence and pointing beyond pragmatic goals—has been introduced to physics with the introduction of the concept of irreversible time into classical thermodynamics and with the transformation and differentiation of this concept within the framework of the theory of dissipative structures. The *essentialization* of time, as temporalized with the means of physics, that lies in this claim is to be confronted in the sequel with the reflexive temporalization of time that has occurred in modern philosophy of time. Finally, a philosophical alternative to Prigogine and Stengers' universalist self-interpretation will be developed on this basis.

## Notes

1. Coveney and Highfield (1990) represents an exception as a work in which the authors pursue in detail the history of modern physical thinking on time.

2. With Clausius the universalization first took place in 1865 (Clausius, 1865, pp. 56ff.). Clausius refers to Thomson's 1852 essay as well as his own comments of 1863 as forerunners.

3. For a survey of the history of the Belousov-Zhabotinsky reaction see: Coveney and Highfield, 1990; Kuhnert and Niedersen, 1987, esp. pp. 40–47. A detailed account is found in Winfree, 1987.

4. See, for example: Coveney and Highfield, 1990, pp. 260–292; Davies, 1988; Ekeland, 1984; Jetschke, 1989; Prigogine, 1997, pp. 89–106; Prigogine, 1998, pp. 35–56; Prigogine and Stengers, 1993, pp. 125–327; .

# PART TWO

# The Reflexive Temporalization of Time in Philosophy

The basic reflexive feature of the temporalization tendency emerging in modern philosophy will be expounded by way of a historical-systematic analysis. This analysis, which begins with Kant and progresses via Husserl and Bergson, will focus on Heidegger's theory of temporality. The reflexive temporalization of philosophy appearing in Heidegger's thinking and the objective temporalization of physics culminating in Prigogine's research will finally be set in relation to one another as two ways of temporalizing time.

## Kant's Theory of Time as the Starting Point of the Reflexive Temporalization Tendency

The transcendental philosophy of time, presented by Kant in the *Transcendental Aesthetic* of his *Critique of Pure Reason* (1781), may be considered the Magna Carta of modern philosophy of time. In the *Critique*, Kant defined time as being "a pure form of sensible intuition" (Kant, 1985, p. 75 [B 47]). There is almost no single philosophical theory which has been misunderstood as often as Kant's definition of time as a pure form of sensible intuition. The standard misinterpretation is that Kant, with his theory, had refuted the reality of time and downgraded it to being a merely subjective illusion. This misunderstanding is widespread not only among philosophers but, above all, among scientists.

The following quote from the British philosopher and founder of analytical philosophy of time, John M. E. McTaggart, provides a significant example of the persistence with which this misunderstanding had established itself within philosophy. In his famous 1908 essay "The Unreality of Time" he writes, "In philosophy, again, time is treated as unreal by Spinoza, by Kant, by and by Schopenhauer" (McTaggart, 1908, p. 457). Scientists such as Albert Einstein or Kurt Gödel, who had read McTaggart (Gödel, 1970, p. 557, footnote), also went along with this prejudice. Thus Gödel, on whose view time had lost its "objective meaning" (Gödel, 1970, p. 557) through the "relativity of simultaneity" (Gödel, 1970, p. 557) proven by Einstein, writes: "In short, it seems that one obtains an unequivocal proof for the view of those philosophers who, like Parmenides, Kant, and the modern idealists, deny the objectivity of change and consider change as an illusion or an appearance due to our special mode of perception" (Gödel, 1970, p. 557). Just as Gödel praises Einstein's work as being physical evidence of the unreality of time allegedly formulated by Kant, McTaggart commends his own work as being an analytic variant of the proof, supposedly demanded by Kant, of time's unreality. In this sense McTaggart writes, "I believe that time is unreal. But I do so for reasons which are not, I think, employed by any of the philosophers whom I have mentioned" (McTaggart, 1908, p. 457).

At this point it would lead too far to examine McTaggart's proof in detail. In summary, however, it may be said that what McTaggart proves is nothing other than what Kant himself had shown long ago: namely not—as McTaggart believed—that time is absolutely unreal, but rather that time has no reality that is independent of the subject. This is an important difference. If time has no subject-independent reality, then this means only that it lacks a certain kind of reality—and not reality altogether. Thus it is not unreal in an indiscriminate sense and a mere illusion. Moreover, having no subject-independent reality is by no means a deficit that devalues the ontological status of time in contrast to the being of other things. For, as Michael Dummett highlighted in the 1960s in his essay "A Defense of McTaggart's Proof of the Unreality of Time" (Dummett, 1960, p. 504), the idea of time as a subject-independent, fully describable reality is in itself a fiction. A fiction presupposing that we have access to a world through which—detached from our finite cognitive conditions—we connect, in the sense of a quasi-divine essential intuition, with inner being.

A decisive flaw in the study of analytic time theory presented by Peter Bieri in the 1970s is that Bieri fails to take account of Dummett's considerations (Bieri, 1972). Bieri's attempt to countercheck McTaggart's proof of the time's unreality with a proof of its reality is marred by this flaw in two respects. First of all Bieri's reading of McTaggart is fixated on those aspects of

McTaggart's approach which might support Bieri's suggestion that McTaggart had attempted to prove the absolute unreality of time. Through this he overlooks those quite central aspects of McTaggart's time theory which, in recourse to James, accentuate the relativistic view of time's reality (McTaggart, 1908, pp. 471ff.). Secondly, from Bieri's one-sided reading of McTaggart there unavoidably results his own attempt, one necessarily destined to fail, to prove time's reality in the strong sense of its being a subject-independent reality that is to be considered " 'real' no matter whether or how we experience it" (Bieri, 1972, p. 11).

Kant had already put an end to this fiction of a subject-independent reality. In contrast to the assumptions made by McTaggart and Gödel in the quotes mentioned here, it was by no means Kant's aim to question the objectiveness of time in the sense of reducing it to the level of "illusion" or "mere appearance." The Kantian linking of time, apprehended with recourse to Newton, Leibniz and scholastic tradition as a subject-independent world structure (Martin, 1955 pp. 11–41), to the transcendental subject is, conversely, far more an attempt to ground time's objectivity in a new, transcendental, manner, one considering the justified doubts expressed by Hume towards the traditional Leibnizian-Newtonian tradition. The point of Kant's reasoning is that time can be ineluctable and *a priori*—that is, generally valid and necessary (Kant, 1985, p. 43f. [B 3f.])—only when it is proven to be an intersubjectively binding condition, and moreover one transculturally and supertemporally valid, for the possibility of knowledge altogether.

Kant, however, highlights sensible intuition as the fundament of all human knowledge—this in contrast to the traditional Platonic concept of knowledge, prevalent through to the time of Leibniz and Newton, according to which only the intelligible can be a true object of knowledge. The basic thesis of the *Critique of Pure Reason* is simultaneously contained in its first sentence. This reads: "In whatever manner and by whatever means a mode of knowledge may relate to objects, intuition is that through which it is in immediate relation to them, and to which all thought as a means is directed" (Kant, 1985, p. 65, [B 33]). It is this basic thesis, that of the primacy of intuition as the basic condition of possibility for all human knowledge, which must be considered in order to understand how far Kant's proof that time is "a pure form of sensible intuition" (Kant, 1985, p. 75 [B 47]) simultaneously assures its objectivity and universality.

Kant's simple thesis, which Gödel—along with most other scientists who have come up against Kant's theory of time—fails to consider, is that all knowledge accessible to humankind—and that includes humankind in our pursual of science (e.g. in physics)—rests on sensible, that is temporal, intuition. Kant thus attempts to secure the objectivity and universality of time precisely through its transcendental subjectivism. This connection is expressed in the following much cited passage from the *Transcendental Aesthetic*. At first it seems that Kant really does want to deny time all reality. He writes: "Time is therefore a purely subjective condition of our (human) intuition (which is always sensible, that is, so far as we are affected by objects), and in itself, apart from the subject, is nothing" (Kant, 1985, p. 77 [B 51]). But the next sentence, usually omitted in citation, is decisive. This states, "Nevertheless, in respect of all appearances, and therefore of all the things which can enter into our experience, it is necessarily objective" (Kant, 1985, p. 78 [B 51]). With this in mind Kant then speaks of the "empirical reality" of time, that is, of its "objective validity in respect of all objects, which allow of ever being given to our senses" (Kant, 1985, p. 78 [B 52]).

Along with the misunderstanding about the unreality of time, the reception of Kant's theory of time is marked by a second consequential shortfall. Strictly speaking, this shortfall is less a misunderstanding than a failure to understand, that is, a narrowed outlook in reception. Decisive aspects of Kant's time theory have long been eclipsed through its equation—one preformed by Schopenhauer (Schopenhauer, 1969, pp. 5–13, 425, 437–452) and Hegel (Hegel, 1974, pp. 432ff.)—with the theory of time in the *Transcendental Aesthetic*. This matter was brought out, using insights gained from Heidegger's book *Kant and the Problem of Metaphysics* (Heidegger, 1990), by Klaus Düsing in his *Examination of Kant's Theory of Time and its Critical Modern Reception* (Düsing, 1980). As Düsing emphasizes at the beginning of his investigation, "Kant's theory of time is, of course, contained only incompletely in the *Transcendental Aesthetic* of the *Critique of Pure Reason*; essential details of this theory are found in the subsequent sections" (Düsing, 1980, p. 2). Similarly, already in §10 of Heidegger's *Kant and the Problem of Metaphysics* we read: "The following interpretation shows how time shifts more and more to the forefront in the course of the individual stages of the laying of the ground for metaphysics, and hereby first reveals its own particular essence in a more original way than the provisional characterization in the Transcendental Aesthetic permits" (Heidegger, 1990, p. 52; cf. also pp. 96, 124).

The obscuring of those aspects of Kant's time theory pointing beyond the *Transcendental Aesthetic* is based on another more profound narrowing of outlook, one no longer dealt with by Düsing. This narrowing of outlook lies in the failure to acknowledge the fact that Kant himself had explicitly relativized his own transcendental universalization of time. Whereas Husserl and Bergson paradigmatically bind the Kantian subjectivization of time back to the finite intentional subject (Husserl) or the living self of pure, flowing duration (Bergson), and simultaneously—against Gödel—stick to the universality of time as a dimension constituting the subjectivity of the subject itself, Kant had already questioned the universality of time which he had initially presupposed.

The decentralization and relativization of the transcendental philosophical time concept hinted at with Kant was consistently carried through by Nietzsche. He spelt out historically the temporality which became noticeable as a blind spot in the rear of the transcendental philosophical treatment of time, one pervading not only all factual knowledge, but the conditions of possibility of knowledge themselves. Thus Nietzsche writes in a fragment from the *Nachlaß* dating from between April and June 1885: "The most established motions of our mind, our regular gymnastics, for example, in presentations of space and time, or in the need for 'justification': this philosophical habitus of the human mind is our actual potency: that is, that in many mental matters we can do nothing else: what one calls psychological necessity. This has become" (Nietzsche, 1980, vol. 11, p. 449, no. 34[89]).

The transition from Kant's transcendental philosophical time theory to Nietzsche's perspectival thinking on time has been brought out by Werner Stegmaier.[1] Stegmaier highlights that the basic problem of modern philosophy of time is "that the representation of time takes place in a time which the representation determines and hence can no longer be represented itself" (Stegmaier, 1987, p. 203). Against the background of this problem he situates Kant and Nietzsche as follows:

> Kant solved it [the basic time-philosophical problem—M.S.] by grasping time as the unrepresentable condition of our representation, that is, as our perspective. The perspective, however, was to be given *a priori*, that is, itself exempted from time. It was Nietzsche who first carried out perspectivization in the philosophy of time too. He gradually drew the perspective of philosophical and scientific knowledge itself into a no longer representable time. (Stegmaier, 1987, p. 204)

This occurs with Nietzsche, Stegmaier continues, "from a historical, aesthetic and practical, temporal representation of time" (Stegmaier, 1987, p. 204). Hence Kant and Nietzsche opened up the horizon for a reflexive temporalization of time which was set out further by Heidegger and is currently being worked on by Derrida (Derrida, 1992), Lyotard (Lyotard, 1991), Rorty (Rorty, 1989, 1994) and others.

Kant's own relativization of the universalization of time that had been carried out in the *Transcendental Aesthetic* is found not in the *Transcendental*

*Aesthetic* itself, but is developed in passing within the framework of the *Transcendental Logic*. The distinction Kant makes in a footnote in the B edition of the *Transcendental Deduction* between time as a "form of intuition" and as "formal intuition" (Kant, 1985, p. 170 [B 160]) is of central importance here (cf. Heidegger, 1976, pp. 294–297; Heidegger, 1977a, pp. 132–139). The distinction to which this footnote relates is already introduced in the content of the main text. The main text reads:

> In the representations of space and time we have *a priori forms* of outer and inner sensible intuition; and to these the synthesis of apprehension of the manifold of appearance must always conform, because in no other way can the synthesis take place at all. But space and time are represented a priori not merely as *forms* of sensible intuition, but as themselves *intuitions* which contain a manifold [of their own], and therefore are represented with the determination of the *unity* of this manifold (*vide* the Transcendental Aesthetic). (Kant, 1985, p. 170 [B 160])

Thus, on Kant's own understanding, the subject of the *Transcendental Aesthetic* is not the form of intuition as such, but a quasi-objective construction: time as formal intuition.

Günter Wohlfart sees this insight of Kant's as being already fixed within the *Transcendental Aesthetic* itself in radicalized form. In his interpretation Wohlfart points to the fact that Kant attempts to prove the "*intuitiveness* of the representation of time" (Wohlfart, 1982, p. 23) in the fourth and fifth arguments in the *Metaphysical Exposition of the Concept of Time*, and in particular to the way he does this. Kant's proof, which demonstrates that time cannot be an intellectual concept, implicitly leaves open the possibility, although not explicitly seized by Kant himself, so Wohlfart continues, of nonetheless apprehending time not as intuition, but as a concept of reason, that is, as an idea: "According to the Kantian proof, space and time might perhaps be not universal, but only singular representations. This means, however, that they could be both (pure) *intuitions* and *ideas*" (Wohlfart, 1982, p. 29).

Kant himself defines the object of the *Transcendental Aesthetic* retrospectively in the previously mentioned footnote to the *Transcendental Logic* as follows: "Space, represented as *object* (as we are required to do in geometry), contains more than mere form of intuition; it also contains *combination* of the manifold, given according to the form of sensibility, in an *intuitive* representation, so that the *form of intuition* gives only a manifold, the *formal intuition* gives unity of representation" (Kant, 1985, p. 170 [B 160f.]). Explicitly referring to the *Transcendental Aesthetic* the annotation continues:

> In the Aesthetic I have treated this unity as belonging merely to sensibility, simply in order to emphasize that it precedes any concept, although, as a matter of fact, it

> presupposes a synthesis which does not belong to the senses but through which all concepts of space and time become possible. For since by its means (in that the understanding determines the sensibility) space and time are first *given* as intuitions, the unity of this *a priori* intuition belongs to space and time, and not to the concept of the understanding. (Kant, 1985, p. 170 [B 161])

This means that the conceived unity of time in the *Transcendental Aesthetic*, which in the "Transcendental Exposition of the Concept of Time" (Kant, 1985, p. 76 [B 48]) at the same time serves as a fundament of the "general doctrine of motion" (Kant, 1985, p. 76 [B 49]), is itself already an objectivized representation of time, that is, one presupposing conceptual or categorial syntheses. It is this uniform linear conception of time, which may be introduced "by analogies" (Kant, 1985, p. 77 [B 50]) in describing "a line progressing to infinity" (Kant, 1985, p. 77 [B 50]), that was universalized and epistemologically legitimized by Kant—not least with regard to Newton's physics—in its "empirical reality" (Kant, 1985, p. 78 [B 52]).

At the same time, however, the second representation of time, appearing behind the back of the objectivized time concept, eludes transcendental philosophical explication. For time, as a form of intuition in the strict sense, forms the horizon, no longer illuminated by Kant, in which time itself can first be dealt with as formal intuition. The universality of the objective time concept in the *Transcendental Aesthetic* is once again decentralized and at the same time methodically relativized by the dimension of this horizon, which itself cannot be accounted for in transcendental philosophical terms. In this sense Heidegger emphasizes in §9 of his *Phenomenological Interpretation of Kant's Critique of Pure Reason* (Heidegger, 1977a), that from Kant's perspective "formal intuition is not a primordial, but a derived representation" (Heidegger, 1977a, p. 132). The way in which Heidegger's analysis of temporality is presented against the background of this insight will be brought out in the sequel.

## Between the Temporalization and Essentialization of Time: Bergson and Husserl

Before taking a detailed look at Heidegger's analysis of temporality, I would like to demonstrate its specifity in demarcation from the contemporaneous time theories of Bergson and Husserl. To this end both Heidegger's analysis of temporality and the time theories of Husserl and Bergson will first of all be situated within the context of debate about time which ran in the first half of the twentieth century.

## Bergson, Husserl and Heidegger in Context

The tension between the universalization and relativization of time, which already dominates the core of Kantian time theory, appears clearly at the center of modern philosophy of time at the beginning of the twentieth century. The context of discussion in the first half of the century extends from the time philosophical positions in a narrow sense of Bergson, Husserl and Heidegger, through a broad field of psychologizing, sociologizing time concepts, or those based on worldview arguments (Hönigswald, Minkowsky, Volkelt, Simmel, Scheler, Klages, Spengler etc.), through to the so-called "philosophy of space and time" (Reichenbach, 1957) whose advocates attempted to transform the philosophical question regarding time into the theoretical scientific examination of relativistic physics.

Characteristic of debate at this time is the fact that, alongside a multitude of authors absolutizing either the aspect of ontological relativity or the aspect of ontological universality, the main exponents in discussion of this period (Bergson, Husserl, Heidegger) themselves argued out the conflict between time relativization and time universalization at the center of their theories. This is ignored by most historical retrospectives dealing with the time debate in the first half of the century. Instead of this, the polar opposition between one-sidedly relativistic and one-sidedly universalistic positions is foregrounded in most surveys. In the following I will refer to the complete accounts by Werner Gent (Gent, 1962, 1965), Charles M. Sherover (Sherover, 1975) and Rudolf Wendorff (Wendorff, 1985).

On the side of the one-sided relativistic positions authors are found who note an undifferentiated and unrelated multitude of particular *Eigenzeiten*. These they abstractly contrast with an opposing unitarily and uniformly conceived world time. On their own self-understanding these authors follow on from the transcendental subjectivization of time carried out by Kant in the *Critique of Pure Reason*'s *Transcendental Aesthetic*. But in their attempts they fall victim to the unreality misunderstanding highlighted above with the example of McTaggart, Einstein and Gödel's Kant reception. With the mediation of the voluntarist time theories of Schopenhauer (cf. Gent, 1962, pp. 229–247; Morin, 1951; Most, 1977, pp. 11–72; Sandbothe, 1989, esp. pp. 159–163) and Schelling (cf. Wieland, 1956) they attempt to radicalize the Kantian subjectivization of time by declaring time to be a variable experiential function of particular subjects (Scheler, Simmel, Spengler, Klages, Hönigswald, Minkowski, Vokelt etc.). By doing this, however, they adhere to the concept of the subject as a timeless authority presupposed by Kant and Schopenhauer (and first questioned by Nietzsche) which is the model according to which the subjectivity of particular subjects is conceived. This distinguishes them from the current advocates of a reflexive temporalization of time in philosophy (Lyotard, Derrida, Rorty).

On the other side of the opposition different variants of time philosophical universalism are found. On the one hand, there are positions following on from Fichte, Schelling and Hegel, which attempt to undermine process-philosophically the Kantian subjectivization of time (Lotze, Whitehead, Alexander) or to reinterpret it logicistically (Cohen, Natorp, Cassirer). On the other hand, positions are found which explicitly and consciously go back before Kant. These include, firstly, the reactivation of pre-Kantian modern theories of time which attempt to derive human experience of time within the empiricist tradition from the successive associative relationship of originally timeless sensations (Mill, Mach, Herbart, Wundt, Russell). Secondly, there are attempts to reactivate pre-modern theories of time which, running counter to the Kantian subjectivization, understand time as being a substantial and subject-independent basic structure of absolute reality (von Baader, Bolzano).

In the surveys mentioned even the positions of those authors who themselves argue out the conflict between both extremes in a differentiated manner in the midst of their theories are usually tailored according to the straightforward opposition of one-sided time relativism and one-sided time universalism. The three outstanding authors to have developed such a higher-level problem formulation in their theories of time are Bergson, Husserl and Heidegger. Common to all three authors is that in their theories of time—unlike, say, Hönigswald (Hönigswald, 1965) or Minkowski (Minkowski, 1970), who may be considered paradigmatic advocates of one-sidedly relativistic time theories—they foreground the deeper dimension of time that is decisive for the constitution of subjectivity itself. In doing this they attempt to illuminate the horizon of time understood (in the sense of the *Transcendental Logic*) as a "form of intuition" (Kant, 1985, p. 170 [B 160]), which although marked out by Kant, was not analyzed by him. The individual experiential time of particular subjects appears from this perspective as a derivative dimension of those underlying structures of temporality which both found the subjectivity of the subject and constitute that which is experienced in experiential time. It is this dimension of a double backward descent that leads Bergson, Husserl and Heidegger to contrast the time relativized by Kant to the transcendental subject's synthetic feats with the universality of a more primordial time through which the subject's subjectivity itself is first constituted.

A few cases from the survey literature might serve as examples of the undifferentiated ordering of these time theories to the one-sidedly relativistic or the one-sidedly universalistic positions. Thus Wendorff ascribes the works of Bergson, Husserl and Heidegger without further delimitation to the entire movement of existential and life-philosophy by equating the basic intention of their approaches with those of Klages, Frobenius, Weininger et al. Common to all of them—Wendorff explains—is the "endeavour to uncover irrational forces," the "distinction between a 'true experiential time' and an 'inauthentic time' contrasting with this," as well as their "grappling with the correct interpretation of the now and its relationship to past and future" (Wendorff, 1985, p. 477). Gent proceeds similarly by presenting Bergson, Husserl and Heidegger, along with Scheler, Spengler, Volkelt and Conrad-Martius, as advocates "of the method of so-called 'direct self-contemplation' . . . [applied to] the experience of time" (Gent, 1965, p. 50f.) whose common goal is the "renunciation of the physical concept of time" (Gent, 1965, p. 58). Whereas Wendorff sees the "philosophical endeavour" of existential and life philosophy's philosophy of time as being centered on "opposing a resolution of the phenomenon of time into a sum of new psychologically explored singular phenomena" (Wendorff, 1985, p. 471)—that is, emphasizes the time-theoretical universalism of existential and life philosophy—Gent highlights instead the relativism of the positions collected by him under the caption "The phenomenological concept of time" and bemoans: "Once again time is repeatedly declassed" (Gent, 1965, p. 58).

Although more differentiated, Sherover still simultaneously proceeds very schematically when he presents Bergson's philosophy of time as one of four attempts "to reassert the independent objectivity of what Kant had termed the two forms of intuition" (Sherover, 1975, p. 157). Bergson appears here alongside Hegel, Lotze and Alexander as representing a "revival of Aristotelian realism" (Sherover, 1975, p. 156) and in this sense as the pronounced advocate of a strong ontological universalism of time. In contrast to this Sherover presents Husserl and Heidegger as representing a weak universality of time documented solely by the temporal structure of human consciousness (Sherover, 1975, pp. 437–465, esp. 438–440). In doing this Sherover foregrounds Husserl's starting point for reflexive questioning of "how we experience our own experience" (Sherover, 1975, p. 438f.). In contrast to him, the time theories of Heidegger and Minkowski initially appear only to be experiments "in existentializing phenomenology" (Sherover, 1975, p. 440). Finally, however, in his chapter on Heidegger, Sherover makes clear that

Heidegger's analysis of temporality sets new standards compared with the time theories of his predecessors: "we have the beginning of an explanation of the integrity of our temporalizing experience *and* of our present attempt to understand it as well" (Sherover, 1975, p. 465).

Indeed, compared with the time theories of Bergson and Husserl, Heidegger's analysis of temporality assumes a particular position. Heidegger's distinction vis-à-vis Bergson and Husserl consists of his developing a set of philosophical instruments, with the help of which the temporality structures constituting subjectivity can be described without being overformed by the traditionally dominant model of linear world time. Heidegger succeeds in this because—unlike Husserl and Bergson, who explicitly adhere to a theoretical understanding of time consciousness in an emphatic sense—he comprehends the temporality of human Dasein in terms of the practical self-projection of human existence.[2] At the same time this also founds the "priority of the future" (Heidegger, 1993, p. 378) which Heidegger brings out with the utmost precision as being the decisive temporal dimension for the practical self-projection of human existence. The time analyses by Bergson and Husserl are also informed by the temporal dimensions (past, present, future) as those basic temporal structures distinguishing human consciousness from linear world time. Yet in contrast to Heidegger they center the fabric of the temporal dimensions not in the future dimension, which is decisive for practical self-projection, but in those dimensions that had been closely linked by a tradition reaching from classical antiquity through to Hegel with theoretical contemplation: the past (Bergson) and the present (Husserl).[3]

Common to both concepts of time—Bergson's model of time centering on the past and Husserl's model of time centering on the present—is that they presuppose time as an inner object accessible to theoretical contemplation: an object revealing itself to phenomenological analysis (Husserl) or to life-philosophical intuition (Bergson) as immediate givenness in the mode of indubitable evidence. This basic theoretical structure, which distinguishes the time conceptions of Husserl and Bergson from Heidegger's temporality, will first be elaborated on for Bergson and for Husserl and then confronted with Heidegger's critical counter design.

## Bergson's Theory of Pure Duration

Bergson had already developed the fundamentals of his philosophy of time in the 1880s. In his study on time entitled *Time and Free Will*[4] he presents "pure duration" (Bergson, 1910, p. 77 and passim) as immediate givenness which is accessible to our original "intuition" (cf. also Bergson, 1913, p. 6) in a distinguished manner and is disguised to us only by the objectifying snares of language, tradition and science. Whereas the quantifying objectification with regard to "material objects" (Bergson, 1910, p. 85), given to us by the spatially ordered sensations of the outer sense, is expedient, quantification with regard to "states of consciousness" (Bergson, 1910, p. 87) seems inappropriate to him. These present themselves to primordial intuition far more as purely temporal and "essentially heterogeneous" (Bergson, 1910, p. 120)—that is organically structured and irreducibly qualitative—structures. But, as Bergson continues, in both our everyday self-interpretation and in science we nonetheless have to contend with a constant tendency to spatialize the pure duration which distinguishes our "deeper self" (Bergson, 1910, p. 125): "Now, let us notice that when we speak of *time*, we generally think of a homogeneous medium in which our conscious states are ranged alongside one another as in space, so as to form a discrete multiplicity" (Bergson, 1910, p. 90).

To counteract this tendency, Bergson considers everyday and "so to speak materialized" (Bergson, 1910, p. 127) time, in which we project the facts of consciousness in analogy with spatial things as being finished, clearly delineable, homogeneous entities, and contrasts this with "duration as quality" (Bergson, 1910, p. 127), in which the constitution of these facts themselves first occurs. In the linear time sequence found in everyday perception we basically do not, according to Bergson, really experience time as time, that is, as "true duration" and "pure succession." In the objectifying apprehension of time of everyday experience the succession assumes far more "the form of a continuous line or a chain" (Bergson, 1910, p. 101). In the imagery of the line we experience a multiplicity of distinct entities as being lined up alongside one another in a simultaneously surveyable order (Bergson, 1910, p. 101f.). Yet it is evident, Bergson continues, "that, in order to perceive a line as a line, it is necessary to take up a position outside it, to take account of the void which surrounds it, and consequently to think of space as three dimensions" (Bergson, 1910, p. 103).

What we describe as "time" in an everyday sense proves to be a spatial construct in Bergson's analysis. Bergson's core proposition is that time is space. With this he attempts to expose not only the classical tradition of occidental philosophy of time, but also the modern tradition going back to Kant (Bergson, 1910, pp. 98ff., 232ff.). In contrast to the conventional so-called—that is "materialized"—time, space proves itself to be the more original notion (Bergson, 1910, p. 98). This, of course, is not Bergson's last word. The very point of his philosophy of time consists of bringing out that space for its part presupposes time as true duration:

> When we explicitly count units by stringing them along a spatial line, it is not the case that, alongside this addition of identical terms standing out from a homogeneous background, an organization of these units is going on in the depths of the soul, a wholly dynamic process . . . without this interpretation and this, so to speak, qualitative process, no addition would be possible. Hence it is through the quality of quantity that we form the idea of quantity without quality. (Bergson, 1910, p. 123)

According to Bergson the dimension of the "present" (Bergson, 1910, p. 227), experience of "simultaneity" (Bergson, 1910, p. 227), which first makes possible the implicitly three-dimensional structure of the spatial time line, is an interlaced fabric of recalled pasts. Bergson evokes the idea of this organic fabric, through which the present is constituted out of the past, in many metaphors in *Time and Free Will.* Yet a detailed analysis does not follow with Bergson. Even the idea that past is that dimension in which true duration is centered,[5] an idea basic to his whole theory of time, is expressed only implicitly in *Time and Free Will.* By contrast, Bergson's prioritization of past is explicitly highlighted in the text *Matter and Memory* that appeared seven years later. Here we read:

> If, on the other hand, what you are considering is the concrete present such as it is actually lived by consciousness, we may say that this present consists, in large measure, in the immediate past. . . . Your perception, however instantaneous, consists then in an incalculable multitude of remembered elements; in truth every perception is already memory. *Practically, we perceive only the past,* the pure present being the invisible progress of the past gnawing into the future. (Bergson, 1988, p. 150)

The immediate past, which permanently presses forward beyond the present into the future and hence absorbs and encroaches on both of these, is identified by Bergson in *Matter and Memory* with the voluntaristic practical and peculiaristic dynamics of the biological life process. All in all the orientation according to "the practical end of all our actual psychical states" (Bergson, 1988, p. 240) attains a positive significance not present in the early work. At the same time, however, in *Matter and Memory* too Bergson sticks to the contemplative theoretical model of true duration which he had outlined in *Time and Free Will* (cf. Flasch, 1993, p. 35f; Giroux, 1971, pp. 43–56; Oger, 1991, pp. xxxiv–xxxvii; Pflug, 1959, pp. 137–197, esp. 162–176). This can be clearly shown with his famous theory of "the two forms of memory" (Bergson, 1988, pp. 79–90), which can be read as a theoretical explication of Proust's *Remembrance of Things Past* (Proust, 1960).

According to Bergson's theory of memory the past divides into two different forms where, although "In normal life they are interpenetrating" (Bergson, 1988, p. 155), simultaneously "each has to abandon some part of its original purity" (Bergson, 1988, p. 155). It is this original purity which Bergson insists on pointedly within the framework of his theory of memory.

Bergson considers as an example the difference between automated, actual presence of the past, expressed in being able to recite a poem from memory, and, in contrast to this, the more virtual and passive recollection of the singular, exactly dateable, phases of the process of learning by heart. Using this example Bergson explains the difference between "true memory" (Bergson, 1988, p. 151), which is "entirely spontaneous" (Bergson, 1988, p. 88) and cognitively "imagines" (Bergson, 1988, p. 82), and the memory based on motoric habit, which "voluntarily" (Bergson, 1988, p. 83) "repeats" (Bergson, 1988, p. 82; cf. pp. 79–90) in action. Only the first memory, separated, as Bergson himself guarantees, by "a difference in kind" (Bergson, 1988, p. 80) from the second, permits descent into the pure sphere of memory. In this we recognize and experience "truly moving in the past" (Bergson, 1988, p. 151) or, as Bergson also puts it, "survival of the past *per se*" (Bergson, 1988, p. 149).

In this sense Bergson highlights at the same time: "there is much more between past and present than a mere difference of degree. My present is that which interests me, which lives for me, and in a word, that which summons me to action; in contrast, my past is essentially powerless" (Bergson, 1988, p. 137). The pure past "is pure from all admixture of sensation, is without attachment to the present, and is, consequently, unextended" (Bergson, 1988, p. 140f.); it is hence distinguished by its "radical powerlessness" (Bergson, 1988, p. 141) as the object of pure contemplation, "regardless of utility or of practical application" (Bergson, 1988, p. 81). Against this background the same problems result for the analysis of "true memory" (Bergson, 1988, p. 151) as outlined by *Matter and Memory* that Bergson had highlighted in individual self-critical reflections for the doctrine of "true duration" in *Time and Free Will*.

Whereas "time as quantity" (Bergson, 1910, p. 129) is the object of the natural sciences (Bergson, 1910, pp. 88ff.), but also that of traditional philosophy (Bergson, 1910, pp. 113, 232ff.) as well as of a "superficial psychology" (Bergson, 1910, p. 139) informed by its basic concepts, the examination of "time as quality" (Bergson, 1910, p. 129)—as Bergson self-critically notes in *Time and Free Will*—has nearly insurmountable difficulties to combat in attempting to establish itself scientifically. The "delicate and fugitive impressions of our individual consciousness" (Bergson, 1910, p. 132), in which "pure duration" (Bergson, 1910, p. 77 and passim) or "pure memory" (Bergson, 1988, p. 137 and passim) articulate themselves, would, in order to be able to enter into competition with the established sciences, have to "struggle on equal terms" (Bergson, 1910, p. 132), which means they would have to "express themselves in precise words" (Bergson, 1910, p. 132). This, however, as Bergson self-critically counters, necessarily leads to a performative contradiction: "these words, as soon as they were formed, would turn against the sensation which gave birth to them, and, invented to show that the sensation is unstable, they would impose on it their own stability" (Bergson, 1910, p. 132).

This, the crux of his own project, which endures from the early study of time through the late work, is made clear by Bergson in *Time and Free Will* using the following example:

> Thus I said that several conscious states are organized into a whole, permeate one another . . . ; thus, by the very language which I was compelled to use, I betrayed the deeply ingrained habit of setting out time in space. From this spatial setting out, already accomplished, we are compelled to borrow the terms which we use to describe the state of mind which has not yet accomplished it: these terms are thus misleading from the very beginning. (Bergson, 1910, pp. 122f.)

Two things are contained in Bergson's self-critical stance. On the one hand, there is the expression of his intellectual honesty, which preserves a certain reserve and scepticism also, particularly even towards his own designs. This aspect of Bergson's thinking, which links it with both the critical impetus of Kantian philosophy and Nietzsche's experimental style of thought, was poignantly accentuated and spelt out in detail by Gilles Deleuze in his deconstructionist study of Bergson (Deleuze, 1988). On the other hand, at the same time a kind of unteachable mysticism comes to light in this Bergsonian stance. In spite of the hitches which his thinking drifts into time and time again, and in spite of the intertwinements between space and time which he himself scrupulously reveals using the concept "endosmosis" (Bergson, 1910, pp. 109, 112, 228) adopted from physics, Bergson sticks to his puritan and basically prospectless program of "ridding" (Bergson, 1910, p. 224) our apprehension of time of the "obsession of the idea of space" (Bergson, 1910, p. 224). This program, which presupposes our having an authentic and indubitably certain access—one bared of all linguistic disguises and everyday objectifications—to the true, that is, eternal and unchanging, essence of time, remained unquestioned by Bergson right through to his late work. It is indicative of the ontological universalism which underlies the core of Bergson's theory of time and comprises the philosophical truth of his romantic invocation of an authentic individuality of pure duration.

## Husserl's Phenomenology of the Consciousness of Internal Time

Husserl's time-philosophical undertaking leads to aporias similar to those which Bergson self-critically identified in his own intellectual approach. This undertaking is documented in the form which was to become most influential in his *Lectures on the Phenomenology of the Consciousness of Internal Time* ("*Lectures*" in the following), held by Husserl in 1905. These *Lectures* were published by Heidegger a year after the appearance of *Being and Time*—although "certainly not out of friendship alone" (Theunissen, 1991, p. 340)—in the edition prepared by Edith Stein. Heidegger obviously made little effort in preparing the lecture manuscript, edited by Edith Stein, and Husserl's notebooks. With this editorially highly problematic publication he had a keen interest in documenting the superiority of his analysis of temporality over the Husserlian theory of time and it only very vaguely follows the original manuscript of Husserl's lectures. Detailed critiques of the editing of the *Lectures* have been provided by Rudolf Boehm (Boehm, 1966, esp. p. xxxviiif.) and Rudolf Bernet (Bernet, 1985, esp. p. lxixf.).[6]

Husserl's starting point is similar to that of Bergson. Renouncing contemporary associationist psychological theories of time (Wundt, Herbart etc.) and critically taking issue with his immediate predecessors (Brentano, Stern) he too initially returns to inner consciousness as the "Suspension of Objective Time" (Husserl, 1991, p. 4). Husserl writes:

> Now when we speak of the analysis of time-consciousness . . . it may indeed seem as if we were already assuming the flow of objective time and then at bottom studying only the subjective conditions of the possibility of an intuition of time and of a proper cognition of time. What we except, however, is not the existence of a world time, the existence of physical duration, and the like, but appearing time, appearing duration, as appearing. These are absolute data that it would be meaningless to doubt. (Husserl, 1991, p. 5)

In this the Cartesian starting point of the analysis is characterized: the basic presupposition that the constitution of inner time consciousness can be explicated with indubitable clearness through phenomenological introspection.

What then proceeds to reveal itself to Husserl's analysis on this basis is "the 'original temporal field' " (Husserl, 1991, p. 6) as "the phenomenological datum through whose empirical apperception the relation to objective time becomes constituted" (Husserl, 1991, p. 7). This original temporal field has its "source-point" (Husserl, 1991, p. 30) in a "primal impression" (Husserl, 1991, p. 371), that is, in the original perception with which "the immanent begins to exist" (Husserl, 1991, p. 29f.). This "absolutely unmodified" (Husserl, 1991, p. 70) thing, which becomes the starting point for the complex fabric of retentions and protentions that Husserl proceeds to unfurl, at the same time characterizes the blind spot, as it were the Parmenidean starting point, of Husserl's entire theory. This theory can be read as the attempt to use the technical tools of phenomenology to grasp in a conceptually precise manner the duration which Bergson had metaphorically described as "a mutual penetration, an interconnexion and organization of elements" (Bergson, 1910, p. 101).

The temporal structure which Bergson circumscribes in ever new approaches as "interpenetration" (Bergson, 1910, p. 123), as "qualitative progress" (Bergson, 1910, p. 123) or as "several conscious states . . . organized into a whole" (Bergson, 1910, p. 122) is grasped by Husserl in his famous time diagram (Husserl, 1991, p. 29) as the "double continuity of running-off modes" (Husserl, 1991, p. 30). Running-off modes or running-off phenomena are those "phenomena that constitute immanent temporal objects" (Husserl, 1991, p. 29), that is, appearances which are themselves constitutive for appearances: "We know that the running-off phenomenon is a continuity of constant changes. This continuity forms an inseparable unity, inseparable into extended sections that could exist by themselves and inseparable into phases that could exist by themselves, into points of the continuity" (Husserl, 1991, p. 29). At the same time, however, in §13 of the *Lectures* as well as in appendix IX (Husserl, 1991, pp. 34ff., 122ff.), Husserl insists that the original temporal field springs from an original perception, an original now, which is to be distinguished strictly from the "comet's tail of retentions" (Husserl, 1991, p. 32) that follow it. This presupposes, however, that the primal impression thus distinguished is a phase somehow delineable from the continuum. Yet how is this to be possible if the "running-off phenomenon" (Husserl, 1991, p. 29) as a whole is simultaneously to constitute "an inseparable unity" (Husserl, 1991, p. 29)?

The answers given by Husserl in the *Lectures* to this pressing question are either circular or speculative in a bad sense. The circular answer reads: "But since primal consciousness and retentions are there, the possibility exists, in reflection, of looking at the constituted experience *and* at the constituting phases, and even of grasping the distinction that obtains, for example, between the original flow as it was intended in the primal consciousness and its retentional modification" (Husserl, 1991, p. 123f.). The following attempted justification is no longer phenomenologically demonstrable, but now merely speculative, and amounts to being a straightforwardly dogmatic imposition: "The primal impression is the absolute beginning of this production, the primal source, that from which everything else is continuously produced. But it itself is not produced; it does not arise as something produced but through *genesis spontanea*; it is primal generation" (Husserl, 1991, p. 106).

The insight first formulated by Bergson in *Matter and Memory*, but then undermined by the ontologization of the past, that "our consciousness of the present is already memory" (Bergson, 1988, p. 151) is on the one hand assumed by Husserl's analysis and is demonstrated in detail in terms of content; yet on the other hand it is explicitly negated and dismissed. Bergson's ontologization of the past corresponds to Husserl's own ontologization of the present. This becomes particularly clear in the following quote, in which Husserl confronts the question of the status of the primal impression:

> We can now pose the question: What about the beginning-phase of an experience that is in the process of becoming constituted? Does it also come to be given only on the basis of retention, and would it be "unconscious if no retention were to follow it? We must say in response to this question: The beginning-phase can become an object only *after* it has elapsed in the indicated way, by means of retention and reflection (or reproduction). But if it were intended *only* by retention, then what confers on it the label "now" would remain incomprehensible. (Husserl, 1991, p. 123)

In the *Lectures*, as edited by Edith Stein and published by Heidegger, Husserl adheres to the last to the transcendental ontological availability of a present which is "by all means characterized in consciousness in quite positive fashion" (Husserl, 1991, p. 123): "Just as the retentional phase is conscious of the preceding phase . . . , so too the primal datum is already intended—specifically, in the original form of the 'now'—without its being something objective. It is precisely this primal consciousness that passes over into retentional modification. . . . If the primal consciousness were not on hand, no retention would even be conceivable" (Husserl, 1991, p. 123).

Husserl's fixation on the primal givenness of an original perceptive presence also underlies his sharp contrasting of primary and secondary memory. Whereas primary memory (retention) goes back to an original presence, secondary memory (reproduction) refers to a non-perception, a non-presence. Retention and reproduction thus stand in opposition to one another as aspects of two acts which are to be radically distinguished. Whereas retention is bound within the original act of "presentation" (*Gegenwärtigung*; Husserl, 1991, p. 40 and passim), reproduction is bound up within the derivative act of "re-presentation" (*Vergegenwärtigung*; Husserl, 1991, p. 42 and passim). With this dislocational strategy Husserl seeks to exteriorize the internal contradiction which pervades "the whole play of primal consciousness and retentions" (Husserl, 1991, p. 123; cf. Derrida, 1993, p. 74).

At the same time he disguises in this way an obvious and plausible alternative to his own approach, one first taken up in the period following by Heidegger, and then extended by Sartre (Sartre, 1956, pp. 107–170; cf. Theunissen, 1991, pp. 131–193) and Merleau-Ponty (Merleau-Ponty, 1962, pp. 410–433; cf. Pieper, 1993). Sartre's criticism of Husserl, found in *Being and Nothingness* (1943), is relevant in this context. Sartre compares Husserl's "protentions" with flies that "batter in vain on the window-panes of the present without shattering them" (Sartre, 1956, p. 109, cf. also p. 100). The alternative developed by Heidegger, Sartre and Merleau-Ponty consists of no longer understanding the internal extension of the original temporal field in terms of the presence of a primal impression which, for itself, is given, and which itself is bound within a linear time sequence constructed according to the model of bracketed world time. The sequence of windows of present, conceived of by Husserl as closed on themselves and statically lined up alongside one another, is resolved by Heidegger, Sartre and Merleau-Ponty into a dynamic referential fabric of temporal dimensions.

This decisive step, first carried out by Heidegger, is the point of transition from a time theory adhering to the ontological universality of time as pure presence to a conception of time which consistently pushes the temporalization of time beyond the limits of presentism. Within these limits Husserl's theory of time might indeed be considered as being that which pushed the temporalization of the present the furthest among classical theories of time. Yet at the same time this means that the reflexive temporalization of time that takes shape with Husserl is to be understood as the temporalization of a present whose ontological universality he preserves in all its might at the foundational level of his examination.

This becomes particularly clear once again in the closing section of the *Lectures*. Husserl defends himself there against a further reaching temporalization of time, one which would radically question the foundations of the presentistic understanding of time and deploying his model of the "self-appearance" (Husserl, 1991, p. 88) of original time flow. The dogmatic imposition of an "absolute subjectivity" (Husserl, 1991, p. 79) and its no longer phenomenologically demonstrable self-givenness was to ensure that "we can no longer speak of a time that belongs to the ultimate constituting consciousness" (Husserl, 1991, p. 83).

An analogous strategy for avoiding the infinite regress which threatens in the conditions of time-theoretical presentism can be found with Bergson. Bergson writes

> If consciousness is aware of anything more than positions, the reason is that it keeps the successive positions in mind and synthesizes them. But how does it carry out a synthesis of this kind? It cannot be by a fresh setting out of these positions in a homogeneous medium, for a fresh synthesis would be necessary to connect the positions with one another, and so on indefinitely. We are thus compelled to admit that we have here to do with a synthesis which is, so to speak, qualitative, a gradual organization of our successive sensations, a unity resembling that of a phrase in a melody. (Bergson, 1910, p. 111)

Not unlike Husserl, Bergson also equates this unity in the closing section of his examination with the, in itself closed, unity of a subjectivity that is utterly transparent to itself. Explicitly distancing himself from the Kantian doctrine of the phenomenal character of knowledge of our selves, Bergson writes: "For if perchance the moments of real duration, perceived by an attentive consciousness, permeated on one another instead of lying side by side . . . , then the self grasped by consciousness would be a free cause, we should have absolute knowledge of ourselves" (Bergson, 1910, p. 235).

In summary, with regard to Husserl and Bergson it is to be highlighted that both presuppose something—be it immanent time of inner time consciousness (Husserl) or true duration of our deeper self (Bergson)—as the object of contemplation in a distinguished theoretical sense. This contemplation of essences is understood according to the model of Cartesian introspection, so that the evident nature of an ineluctable fact is to present itself within the midst of subjectivity: an instance of supposedly genuinely philosophical evidence, which might be questioned by no empirical science, by no contingent change that might occur in the history of humanity, its institutions and habits.

This claim necessarily leads to the margins of language and to the limits of its possibilities of expression. This becomes clear not only, as already shown, with Bergson, but also comes to bear with Husserl. Thus in the almost resigned §39 of Husserl's *Lectures*—already drawn upon above—bearing the title "The Time-constituting Flow As Absolute Subjectivity" we read:

> We can say nothing other than the following: This flow is something we speak of *in conformity with what is constituted*, but it is not "something in objective time." It

is *absolute subjectivity* and has the absolute properties of something to be designated *metaphorically* as " 'low'; of something that originates in a point of actuality, in a primal source-point, 'the now,' and so on. In the actuality-experience we have the primal source-point and a continuity of moments of reverberation. For all of this we lack names." (Husserl, 1991, p. 79)

If one bears in mind that it is exactly this question of the connection between temporality and subjectivity that stands in the center of *Being and Time*, then ultimately it becomes clear once again that Heidegger was able to present Husserl's lectures on time, which Heidegger edited a year after the appearance of *Being and Time*, as being an excellent illustration of the necessity and the strength of his own undertaking.

## The Reflexive Temporalization of Time in Heidegger's Analysis of Temporality

Heidegger unfurled his analysis of temporality in the second division of the first part of *Being and Time* (1927). It is helpful to distinguish two things with regard to Heidegger's early, still fragmentary, master work: the complete undertaking of a fundamental ontology, which was suggested, but not realized; and the analysis of Dasein, which was actually carried out. I would like first to develop this distinction, which underlies the following interpretation, in order then to determine the systematic position of the time problem within *Being and Time* on the basis of this and to outline a pragmatic interpretation of Heidegger's analysis of temporality.

### Fundamental Ontology and the Analysis of Dasein: The Pragmatic Approach to Its Interpretation

Heidegger's fundamental ontological starting point is the "crisis in . . . 'foundations' " (Heidegger, 1993, p. 29) of modern science diagnosed in the introduction to *Being and Time*. Heidegger takes this as providing occasion to drive the undertaking of philosophically "laying the foundations for the sciences" (Heidegger, 1993, p. 30), that going back to Plato and Aristotle and transcendental philosophically radicalized by Kant, towards a further—supposedly ultimate and decisive—foundational step. The by no means self-evident assumption linking this project with ancient ontology and modern transcendental philosophy is formulated by Heidegger as follows: "Such research must run ahead of the positive sciences, and it *can*" (Heidegger, 1993, p. 30).

Philosophy is understood here—more radically still than with the early Wittgenstein or in logical empiricism—as being the distinguished science which as "productive logic . . . leaps ahead, as it were, into some area of Being, discloses it for the first time in the constitution of its Being, and, after thus arriving at the structures within it, makes these available to the positive sciences as transparent assignments for their inquiry" (Heidegger, 1993, p. 30f.). This program of foundation, Heidegger continues, has always been the task of "ontology taken in the widest sense" (Heidegger, 1993, p. 31), and is questioned by modern science's foundational crisis only with respect to its radicality, but not with respect to its fundamental necessity: "Ontological inquiry is indeed more primordial, as over the ontical inquiry of the positive sciences. But it remains itself naïve and opaque if in its researches into the Being of entities it fails to discuss the meaning of Being in general" (Heidegger, 1993, p. 31). Hence, as Heidegger concludes, fundamental ontology is required as the genuinely philosophical attempt to reveal the conditions "for the possibility of those ontologies themselves which are prior to the ontical sciences and which provide their foundations" (Heidegger, 1993, p. 31).

This claim must appear anachronistic in view of the factual situation of the individual empirical sciences, which in the nineteenth and twentieth centuries had become ever more differentiated, respectively developing their own methods of demonstration, traditions and institutions. This has been demonstrated by Herbert Schnädelbach in his book *Philosophy in Germany 1831–1933* (Schnädelbach, 1983) by looking at the "structural change" (Schnädelbach, 1983, p. 94) and the "energization of science" (Schnädelbach, 1983, p. 106) which occurred in the nineteenth and early twentieth centuries. In his study he brings out the way in which the classical philosophical concept of science, centering on generality, necessity and truth, was replaced in this period by a new kind of scientific practice. Within the framework of this new scientific practice, one, however, not sufficiently reflected on at the level of philosophy of science, it is "the rules of procedure acknowledged by physicists themselves which now define the scientific character of the sciences" (Schnädelbach, 1983, p. 109). Amidst the spectrum of reactions to the "identity crisis of philosophy" (Schnädelbach, 1983, p. 119) triggered by this, Heidegger's fundamental ontology appears to Schnädelbach to be an attempted "rehabilitation of philosophy as metaphysics" (Schnädelbach, 1983, p. 137). An attempt, Schnädelbach continues, understanding itself as a "counter-attack against the occupation of reality by the empirical sciences" (Schnädelbach, 1983, p. 233).

Indeed there are passages with Heidegger in which he lays himself open to this accusation. For example, at the end of his lectures on *The Basic Problems of Phenomenology*, held in Marburg in the 1927 Summer Semester, he writes: "We confront the task . . . of inquiring even beyond being as to that upon which being itself, as being, is projected. This seems to be a curious enterprise, to inquire beyond being; perhaps it has arisen from the fatal embarrassment that the problems have emanated from philosophy; it is apparently merely the despairing attempt of philosophy to assert itself as over against the so-called facts" (Heidegger, 1982, p. 282). This attempt expresses a continuation of the Cartesian insistence on philosophical justification, methodical certainty and science's ontological universality. However, the route in fact taken by Heidegger in the first division of *Being and Time* in order to reveal the "ontico-ontological condition for the possibility of any ontologies" (Heidegger, 1993, p. 34) simultaneously points in another direction. With this I come to the hermeneutic difference existing between the original complete outline of a theoretical fundamental ontology and the factual realization of the pragmatic analysis of Dasein.

When in the introduction to *Being and Time* Heidegger distinguishes human Dasein as the "primary example to be *interrogated* in the question of Being" (Heidegger, 1993, p. 28) and hence as the methodological outset of fundamental ontology, it looks to begin with like a simple reprise of the transcendental philosophical turn which Kant had given to traditional ontology. Yet already the first, still completely formal, definition that Heidegger provides of what he calls "Dasein" (Heidegger, 1993, p. 26 and passim)—what in the tradition traded under the name "subject" or "consciousness"[7]—shows that he understands the theme of "Dasein" as being something more and something other than transcendental subjectivity. Heidegger writes: "Dasein . . . is ontologically distinguished by the fact that, in its very Being, that Being is an *issue* for it" (Heidegger, 1993, p. 32). In this definition lies a threefold demarcation from the tradition of philosophical theory.

The first demarcation, directed immediately against Kant, consists in that Heidegger understands the structure of behavior towards oneself, which Kant apprehends as a reflexive structure of self-consciousness, as being not simply a relation in which Dasein stands to itself, but one in which it is concerned with its Being. By italicizing "issue" (Heidegger, 1993, p. 32) Heidegger makes clear where the difference lies. Dasein stands in relation to itself not as to a finished available object, that is one which is "present-at-hand"; rather, in its behavior towards itself, it is an issue for itself as an entity (*Seiendes*) that is essentially open to the future and which respectively projects its own possibilities: "That kind of Being towards which Dasein can comport itself in one way or another, and always does comport itself somehow, we call *existence*" (Heidegger, 1993, p. 32). This means that Dasein refers to itself not in an abstract theoretical manner through a merely cognitive relation as something objectively present-at-hand, but in a concrete practical manner as its own future Being, that is, as the individual existence which is to be respectively carried out and pragmatically shaped.

The second demarcation builds directly on this. Its point is directed—as Tugendhat has shown in the Heidegger passages of his lectures *Self-Consciousness and Self-Determination*—against Aristotle, who in his *Nicomachean Ethics* had himself already put "nontheoretical, practical relation to one's own being" (Tugendhat, 1986, p. 158) in the foreground. Heidegger's demarcation from Aristotle consists of his highlighting that Dasein is concerned with its Being *in its Being*. Dasein's practical relationship to Being is not an active striving, which it can either engage in or not, rather it is a passive occurrence, into which Dasein is thrown and is forced to engage in. In this sense Heidegger highlights that Dasein "in each case . . . has its Being to be, and has it as its own" (Heidegger, 1993, p. 32f.). And Tugendhat, who has pointedly brought out "that man has his being *to be*" (Tugendhat, 1986, p. 158), explains: "I face a range of decisions as to which way I want to carry out my being, but the fact that I have to carry it out is given to me" (Tugendhat, 1986, p. 158f.).

The third demarcation from tradition has likewise been highlighted by Tugendhat. It consists of the fact that Heidegger "attempts for the first time to extract ontological capital from this phenomenon" (Tugendhat, 1986, p. 159). For, whereas Aristotle and the tradition drew no consequences for ontology from the possibility of practical reference to Being, Heidegger, Tugendhat continues, attempted to show that the "Being" which is to be engaged in has a sense quite different to that of "Being" as presentness-at-hand which is to be observed. According to Heidegger's entire conception, the inner intersection between analysis of Dasein and fundamental ontology lies in this third point of demarcation.

Tugendhat himself highlights this intersection by distinguishing in his interpretation between a stronger and a weaker Heideggerian thesis: "I mean by the weaker thesis the contention that the meaning of being differs in accordance with whether it is taken theoretically, as something that is asserted, or practically in the previously specified sense, as something to be carried out. The stronger thesis is that *being* in the sense of presence-at-hand is not only not the only sense of *being*, but also a sense of *being* that is derivative in contrast to that of the to-be" (Tugendhat, 1986, p. 160). For Tugendhat the weaker thesis proves to be the theoretically more fruitful insight of the analysis of Dasein, which can be made plausible by means of linguistic analysis. The strong thesis, however, which "cannot be grasped in language" (Tugendhat, 1986, p. 166) insofar as it presupposes the distinction between "readiness-to-hand" (*Zuhandenheit*) and "presentness-at-hand" (*Vorhandenheit*; cf. Heidegger, 1993, §§15–18, pp. 95–122), is rejected by Tugendhat as being "speculative in the sense that one cannot specify which criteria are to be relevant in evaluating its correctness" (Tugendhat, 1986, p. 166).

The productive point of the analysis of Dasein—as is demonstrated by the three demarcation points developed here—lies in Heidegger's uncovering of the genuinely practical sense of human Dasein's Being. This uncovering takes place, firstly, in the temporalization of behavior towards oneself as a relation toward one's own to-be, that is, to one's own respective future. It takes place, secondly, as the negation of behavior towards oneself as being a relation which the human subject engages in not of its own account, but into which it is thrust. And it takes place, thirdly, as the ontologization of behavior towards oneself as a relation which, in its temporality and negativity, attains an independent character of Being to be distinguished from that of the non-Dasein-like entity encountered in our everyday dealings with the world as readiness-to-hand or presentness-at-hand. This last and decisive step made by Heidegger's analysis of Dasein contains both the foundation for his analysis of temporality and the transition to the theoretically directed programme of fundamental ontology.

However, the distinction introduced by Tugendhat between a weak and strong Heideggerian thesis ought to be extended with a further decisive point. On top of the strong thesis in Tugendhat's sense comes a still stronger thesis of Heidegger's. It consists of the assumption—obtained with the guidance of the analysis of human Dasein—that not only is the practical sense of Being the more fundamental as opposed to the theoretical sense of Being, but also, because of its fundamentality, that it can be transferred to the entire world of non-Dasein-like entities. It is this stronger thesis which effects the actual lapse from Daseins-analytical pragmatism to fundamental-ontological theoreticism. Although this thesis was prepared for by the foundational thesis the two do not coincide.

Heidegger's strong thesis (in Tugendhat's sense) should be defended against Tugendhat. For the pragmatic point of Heidegger's existential conception of science lies precisely in the thesis that the theoretical sense of Being is founded in the practical sense of Being. This is a point which the analytic philosopher Tugendhat cannot go along with, because he himself adheres to the theoretical project of a "formal science, which, in the shape of a formal semantics, underlies all sciences" (Tugendhat, 1982, p. 31).[8] The analytic philosophical argument he adduces against Heidegger's thesis—that the theoretical scientific sense of Being is founded in the technical practical sense of Being—is itself an argument which presupposes for its part that the formal semantic structure of our language be considered an *a priori* reference and the distinguished foundational ground of science. The very idea, however, that there be such a final theoretical reference is questioned by Heidegger's stronger thesis. Hence, Tugendhat's argument cannot be considered proof of the untenability of Heidegger's thesis, for it does not entertain this thesis at all first, but instead presupposes its untenability through the recourse to analytic linguistic reference.

The stronger thesis of Heidegger's, which must be designated as speculative not only from the perspective of analytic philosophy, but also from the viewpoint of a theory of science founded on Dasein analysis, was first dealt with by Tugendhat in a later essay. There he decouples the "primordiality thesis" (Tugendhat, 1992a, p. 131), that is, the assumption—which in *Self-consciousness and Self-determination* was called the strong thesis—of the pragmatic foundedness of the theoretical sense of Being, from consideration of the "thesis planned for the third division [of *Being and Time*]" (Tugendhat, 1992a, p. 131), according to which Dasein's practical sense of Being can also be transferred to non-Dasein-like entities and hence promoted to being the absolute sense of Being.

However, in this case too, in which his criticism is justified, Tugendhat's argument cannot convince. It reads: "The transfer of a structure which in essence is consciousness—or Dasein-like to anything else—and even to Being—makes no sense" (Tugendhat, 1992a, p. 132). Here too Tugendhat's argument presupposes a prior analytic philosophical criterion of meaningfulness which seems to permit judgement of Heidegger's thesis by way of a process of theoretical examination. Against this it is to be objected, from the perspective of a pragmatically founded understanding of theory, that the sense of such a transfer cannot be judged *a priori*, but can be demonstrated only in practical corroboration and in being spelt out in concrete terms of particular sciences. Admittedly, it is just this that Heidegger neglects to do by preferring a purely theoretical attempt at justification, one which for its part thinks itself capable of using a fundamental ontological interpretation of the problem of temporality to attain to an *a priori* foundation of the practical sense of Being as the sense of Being altogether.

The following pragmatic reading of Heidegger's analysis of temporality interprets the second division of *Being and Time*, bearing the title "Dasein and Temporality," from the context of the factually realized work. This means that it comprehends the analysis of temporality in terms of the basic pragmatic trait of the "prepatory fundamental analysis of Dasein" (Heidegger, 1993, pp. 65ff.) carried out by Heidegger in the first division of *Being and Time*. The interpretation hence abstracts as far as possible from the overall fundamental ontological perspective, which although outlined by Heidegger in the introduction, was itself in fact never realized. The framework of *Being and Time*'s fundamental ontological project will be drawn upon only insofar as it directly inscribed itself within the analysis of temporality.

The foundations of a pragmatic Heidegger interpretation have been laid by Robert Brandom (Brandom, 1983), Richard Rorty (Rorty, 1984, 1991) and Mark Okrent (Okrent, 1988) in the USA, as well as by Ernst Tugendhat (Tugendhat, 1970, 1979, 1992a, 1992b), Karl-Otto Apel (Apel, 1973, vol. 1, pp. 225–334; Apel, 1991) and Carl Friedrich Gethmann (Gethmann, 1974, 1993) in Germany. In their interpretations the named authors have concentrated on the first division of *Being and Time* and have more or less excluded the analysis of temporality developed by Heidegger in the second division of *Being and Time*. The interpretative reticence of the pragmatic Heidegger interpreters with regard to the analysis of temporality results from the position assumed by the analysis of temporality within the systematic architectonics of *Being and Time*. In the architectonics of *Being and Time* the analysis of temporality establishes the transition from the analysis of Dasein to the overall project of a fundamental ontology, comprehending Being and Dasein, which Heidegger no longer realized. For this reason, the attempt to pave the way for a pragmatic reading of Heidegger's analysis of temporality requires further methodical preparation. The position of the problem of time within the complete fundamental ontological project, outlined by Heidegger in the introduction to *Being and Time*, must be explained. In doing this the aim is simultaneously to take a more detailed look at the overall conception of *Being and Time*, which reaches beyond the work as realized, to help in demarcation.

## On the Systematic Position of the Problem of Time within the Architectonics of *Being and Time*

The title "Being and Time" already provides a double clue as to the position assumed by the problem of time in *Being and Time*. First of all, by adopting the time concept in the title of the book Heidegger is pointing out that the problem of time is of central importance to his entire project. Secondly, the order in which "Being" and "time" appear indicates that the question regarding time is afforded second place to the question of Being. The original overall conception of *Being and Time* is developed by Heidegger in his introduction under the title "Exposition of the Question of the Meaning of Being" (Heidegger, 1993, p. 22). It is preceded by a quote from Plato's *Sophist* which Heidegger makes the starting point in highlighting the basic intention and the provisional aim of *Being and Time*. Heidegger's first sentence in *Being and Time*, in which he comments on the Plato quote, unfurls the "question of what we really mean by the word 'being' " (Heidegger, 1993, p. 19) as being the germ-cell of the treatise: "So it is fitting that we should raise anew *the question of the meaning of Being*" (Heidegger, 1993, p. 19). The way in which the basic question of being is systematically taken into view, as expressed in the quote and extensively set out by Heidegger in the introduction, confirms in principle the thesis of the priority of the question of Being over the question of time which results from the order of "Being" and "time" in the title.

Explicit confirmation of the thesis is provided by Heidegger when he directly deals with the relationship in which the question of Being stands to the problem of time. In an short intermediate step he had pointed out beforehand that "we in our time" (Heidegger, 1993, p. 19) not only no longer know what we actually mean with the term "being," but that, furthermore, this not-knowing, the worthiness of asking the question about being, has drifted into oblivion. This intermediate step is important for Heidegger's argument insofar as the "aim in the following treatise" (Heidegger, 1993, p. 19) results from it. This aim does not consist, according to Heidegger, of providing an answer to the question of Being. Its aim is rather to "reawaken an understanding for the meaning of this question" (Heidegger, 1993, p. 19).

Against the background of this seemingly reserved intention—no new answers are sought, rather an old question is to be reawakened—Heidegger describes in his next step the relationship in which the question of Being and the problem of time stand to one another. He writes: the overall intention "is to work out the question of the meaning of *Being* and to do so concretely. Our provisional aim is the Interpretation of *time* as the possible horizon for any understanding whatsoever of Being" (Heidegger, 1993, p. 19). The two sentences identify the conceptual precedence attributed to the question of Being by Heidegger over the question of time at the formal level of arrangement. The fundamental intention of *Being and Time* appears to be to work out the question of the meaning of "Being." Conversely the interpretation of time is declared to be the "provisional aim" (Heidegger, 1993, p. 19) and is hence incorporated as a stopover in the overall fundamental ontological undertaking, one merely to be methodically steered towards in advance, but which is in no way decisive for the matter at hand.

Up to this point the relationship between Being and time determined by

Heidegger's overall conception seems to be in agreement with the vague clues to be derived from the order of the two concepts in the title. The problems begin, however, when one subjects the content of the two quoted sentences to a more exacting interpretation. If time—as stated in the passage drawn upon—is to be identified as the "possible horizon for any understanding whatsoever of Being" (Heidegger, 1993, p. 19), is time then not set as being the more comprehensive dimension? And does this not mean that, precisely from Heidegger's perspective, one cannot speak of a priority of Being over time, but must emphasize far more the precedence of the time understood as the horizon for Being instead?

In fact the early Heidegger's overall fundamental ontological program consists of universalizing time in such a way that the constitution of what the philosophical tradition had always opposed with time as something eternal and unchanging—i.e. Being—itself now becomes obviously temporal. This claim is made explicit by Heidegger in paragraphs 11–14 of §5 of *Being and Time*. In these passages, which are decisive in determining the position of the problem of time in the overall conception of *Being and Time*, we read: "If Being is to be conceived in terms of time, and if, indeed, its various modes and derivatives are to become intelligible in their respective modifications and derivations by taking time into consideration, then Being itself . . . is thus made visible in its 'temporal' character" (Heidegger, 1993, p. 40).

If this claim is taken literally then the apparent reserve exhibited by Heidegger in his preface is exposed as glossy rhetoric. Against Heidegger's initial assurance, *Being and Time* is obviously concerned not only with making headway in "work[ing] out the question of the meaning of *Being* and [doing] so concretely" (Heidegger, 1993, p. 19). The intention of the book is rather to demonstrate the plausibility of an "answer to the question of Being" (Heidegger, 1993, p. 40) that is presupposed from the start. This answer reads: the meaning of Being is time.

Summing up against the background of similar considerations Tugendhat highlights in his essay "Heidegger's Question of Being": "So I come to the result that Heidegger had already posed his *question* of Being in a form . . . , which *only* yields an understandable meaning when one already anticipates the answer that time is the meaning of Being. Thus Heidegger had formulated his question in such a way that it has no meaning independently of the answer he already had in mind" (Tugendhat, 1992a, p. 113f.). And Tugendhat adds: "Something of Heidegger's style of thought is demonstrated in this. Had he been made aware of this connection, he would still have been able to extract from it a particular profundity" (Tugendhat, 1992a, p. 114).

Against the backdrop of the anticipated answer the precedence of the question of Being over the problem of time proves to be superficial. For it seems to be undermined by a more profound primacy of time, within which the question of Being takes shape. Of course, it should be highlighted immediately that what at first glance might appear to be a contradiction and aporia in the conception of *Being and Time*, can be easily defused using the distinction between the material and methodical aspect of the relationship between Being and time. The thesis that the sense of Being is time in no way casts doubt on the material primacy of the question of Being for the overall fundamental ontological concept of *Being and Time*.

This is demonstrated, for instance, when Heidegger, precisely while pointing out the methodical significance of the time problem, at the same time accentuates the material priority of the question of Being. Thus in §5 he writes: "the fundamental ontological task of Interpreting Being as such includes working out the *Temporality of Being*" (Heidegger, 1993, p. 40). In a manner similar to the sentences quoted above from the prelude to *Being and Time*, in which the "Interpretation of *time*" appears as the "provisional" (Heidegger, 1993, p. 19), but not the final and decisive aim of *Being and Time*, the working out of the temporality of Being is here also presented as an aspect of a comprehensive treatment of the question of Being. This clearly shows that in the original overall conception of *Being and Time* time was not aimed at as an independent theme, but that the theory of time was conceptionally incorporated in the fundamental ontological perspective of inquiry from which Heidegger had set out when developing his approach in the introduction.

Simultaneously this material precedence of the question of Being over the problem of time provides evidence of what Heidegger highlights in §2 in his hermeneutic analysis of "what belongs to any question whatsoever" (Heidegger, 1993, p. 24): "Inquiry, as a kind of seeking, must be guided beforehand by what is sought. So the meaning of Being must already be available to us in some way" (Heidegger, 1993, p. 24). Posing the question concerning Being in a suitable manner means, for Heidegger, revealing the pattern that philosophical tradition had always used to provide the answer to this question. Heidegger's thesis is that the Greeks had already conceived of Being within the horizon of time—albeit without having accounted for this to themselves. Heidegger writes on this: "The outward evidence for this (although it is *merely* outward evidence) is the treatment of the meaning of Being as παρουσία or ουσια, which signifies, in ontologico-Temporal terms, 'presence' ['Anwesenheit']. Entities are grasped in their Being as 'presence'; this means that they are understood with regard to a definite mode of time—the '*Present*'" (Heidegger, 1993, p. 47). The methodical precedence of the problem of time is not a central theme of *Being and Time* in its own right from the start, rather it becomes this only because the central material question concerning Being methodically points to time as its answer.

But how did Heidegger intend to go about producing the connection between Being and time, which from the perspective of tradition was all but self-evident? What was the link which Heidegger thought would allow him to present time as the answer to the question of Being? In *Being and Time* Heidegger thought he could identify this link by uncovering temporality as Dasein's basic constitution, which was to correspond to the temporality of Being, lying at the core of non-Dasein-like entities, that he sought to reveal. The basic thesis on which the original overall conception of *Being and Time* is based states that two things can be demonstrated on the ground of an existential analysis of Dasein: firstly, that the meaning of Dasein, that is the Being of consciousness, is temporality; and secondly, that a direct route leads from the temporality of Dasein to the temporality of Being. According to the original concept of *Being and Time*—introduced by Heidegger in his "Design of the Treatise" (Heidegger, 1993, pp. 63ff.) in §8—the latter was to be demonstrated by way of a time-theoretical "destruction of the history of ontology" (Heidegger, 1993, p. 63). Through the destruction of traditional ontology, which wrongly understood itself as being time-neutral, the hidden temporal pattern underlying the various historical answers philosophy had given to the question of Being was to be uncovered in the second part of *Being and Time*.

Neither the planned second part of the work, nor even that division of the first part of *Being and Time* in which, under the heading "Time and Being," the transition from time to Being was to be carried out were realized by Heidegger. To this extent *Being and Time* remained a fragmentary work in a two respects. A comparison of the existing work's table of contents with the already cited "Design of the Treatise" (Heidegger, 1993, p. 63) of §8 makes this clear. *Being and Time* was to have two main parts. The title of the first part, which was to consist of three divisions, reads: "the Interpretation of Dasein in terms of temporality, and the explication of time as the transcendental horizon for the question of Being" (Heidegger, 1993, p. 63). The planned title of the second part was: "basic features of a phenomenological destruction of the history of ontology, with the problematic of Temporality as our clue" (Heidegger, 1993, p. 63). Heidegger in fact wrote only the first two divisions of the first part. The third division of the first part and the entire second part were never realized by Heidegger. Some of the basic ideas of the third division of the first part are found in Heidegger's lecture *The Basic Problems of Phenomenology* (Heidegger, 1982).

The dual fragmentary character of *Being and Time* leads to the fact that the question of time dropped out in two regards in the realized work. The analysis of Dasein (part 1, division 1) was indeed interpreted in terms of temporality (part 1, division 2), but the temporally understood Dasein was no longer interpreted explicitly in terms of the question of Being (part 1, the missing division 3). Not to mention the working out of the question of Being itself that was to follow in the historical perspective on the basis of the clarifications gained through the analysis of Dasein (missing part 2). What remains is the "Interpretation of Dasein in terms of temporality" (Heidegger, 1993, p. 63).

The central difficulty with which a pragmatic reading of *Being and Time* sees itself faced lies in that the analysis of temporality actually carried out by Heidegger represents a patchwork: this consists on the one hand of independent and tenable analyses relating to phenomena; on the other hand, however, fundamental ontological mantles, anticipations and constrictions of outlook enter in which are determined by the unrealized finale of *Being and Time*'s original overall conception.[9] The following account attempts first to bring out the phenomenological strength and material tenability of Heidegger's analysis of temporality as delineated from the time theories of Bergson and Husserl and then, against this background, to demonstrate critically its inner intertwinement with the overall fundamental ontological project.

### Pragmatic Interpretation of Heidegger's Analysis of Temporality

Unlike Husserl and Bergson, who did not directly relate their theories of time to Kant, Heidegger's early thinking develops by taking issue directly with Kant. This was clearly expressed in his lecture *Phenomenological Interpretation of Kant's Critique of Pure Reason*, held in Marburg in the year of *Being and Time*'s publication, as well as in the book *Kant and the Problem of Metaphysics*, published in 1929, and in the references to Kant found in *Being and Time* itself (cf. also Sherover, 1971; Düsing, 1992). In directly taking issue with Kant, Heidegger was led to break with the theoretical kind of approach to the problem of time that had already determined the *Critique of Pure Reason* and been retained by Bergson and Husserl. The question of time as a pure form of sensible intuition—which is left open by Kant and reformulated by Bergson and Husserl as the question of the temporality inherent in subjectivity—becomes with Heidegger a question of the genuinely practical mode of the temporal self-design of human existence.

"Dasein" is Heidegger's term for what is called "subject" or "I think" with Kant. The direct dispute with Kant which pervades Heidegger's early main work is already expressed in this basic concept of *Being and Time*. Heidegger takes the view that Kant, by fixing it as "I think," reduces the transcendental subject to the aspect of theoretical knowledge. According to Heidegger, the human is not a being that aims first and foremost to cognize the present-at-hand (*Vorhandene*). As Dasein, it is far more a being which has always been thrust into its "there" (*Da*), and thus did not first begin, artificially and retrospectively, to construct a cognitive relationship to the outer world, but rather one which had always finds itself practically related to its concrete environment—to the "ready-to-hand" (*Zuhandene*; Heidegger, 1993, p. 98). In this sense Heidegger highlights against Kant: "The 'I' is not just an 'I think,' but an 'I think something' " (Heidegger, 1993, p. 367). And he explains: "Kant has indeed avoided cutting the 'I' adrift from thinking; but he has done so without starting with the 'I think' itself in its full essential content as an 'I think something,' and above all, without seeing what is ontologically 'presupposed' in taking the 'I think something' as a basic characteristic of the self" (Heidegger, 1993, p. 367). This postulate is the "Being-in-the-world" of Dasein. Since, however, Kant "did not see the phenomenon of the world" (Heidegger, 1993, p. 368), Heidegger's basic insight must remain hidden to him: "In saying 'I,' Dasein expresses itself as Being-in-the-world" (Heidegger, 1993, p. 368).

Indeed, like Kant, Heidegger also asks about conditions of possibility. For him, however, it is not an abstract matter concerning the possible conditions of knowledge, but quite concretely concerning the conditions of possibility of our Being-in-the-world. In the second division of *Being and Time* Heidegger uncovers "temporality" (Heidegger, 1993, p. 274 and passim) as being the fundamental dimension underlying Dasein's structure of care (*Sorge*), which he had brought out in the first division of *Being and Time*. With recourse to Kierkegaard, he describes the "double-movement" (Kierkegaard, 1987, pp. 36, 119) which brings Dasein to its *Da* ("there"), hence opening it for itself and to the world as a double temporal occurrence.

The first partial movement in this occurrence consists of the anticipation (*Vorlaufen*) of the future, and the second partial movement in coming back to the present as an openness for the world being encountered that is determined by the past—or, as Heidegger puts it, the "having been" (Heidegger, 1993, p. 373). In summary Heidegger writes: "Coming back to itself futurally, resoluteness brings itself into the Situation by making present. The character of 'having been' arises from the future, and in such a way that the future 'has been' (or better, which is 'in the process of having been') releases from itself the present. This phenomenon has the unity of a future which makes present in the process of having been; we designate it as '*temporality*'" (Heidegger,

1993, p. 374). At the existential level of conditions of possibility, the concern here is not the concrete future, determined by certain substantive aims, but the future in general, of which we read: "By the term 'futural,' we do not here have in view a 'now' which has *not yet* become 'actual' and which sometime *will be* for the first time. We have in view the coming [Kunft] in which Dasein, in its ownmost potentiality-for-Being, comes towards itself" (Heidegger, 1993, p. 373).

Heidegger's designation of this ecstatic basic structure of Dasein as "transcendence" (Heidegger, 1993, pp. 62, 414ff. and passim) has also given cause to infer theological implications here. Yet Heidegger defended himself against such a reading of his work from an early stage. Already in his early lecture "The Concept of Time" to theologians in Marburg in 1924, in which he had just formulated the general ideas behind his analysis of temporality for the first time, he emphasizes, completely in the spirit of Kant: "The philosopher does not believe. If the philosopher asks about time, then he has resolved *to understand time in terms of time*" (Heidegger, 1992, p. 1f.). To understand time in terms of time means thinking about time temporally, or to be in favor of a temporalization of time. In this sense, with *Being and Time* in mind, Rorty writes: "Heidegger would like to recapture a sense of what time was like before it fell under the spell of eternity, what we were before we became obsessed by the need for an overarching context which would subsume and explain us" (Rorty, 1991, p. 34). Such is the thoroughly secular nature of Heidegger's program, and it is against this background that his definition of "future" (*Zukunft*) as being the "coming [Kunft] in which Dasein, in its ownmost potentiality-for-Being, comes towards itself" (Heidegger, 1993, p. 373) is to be understood.

According to Heidegger the reason that our "Ways of Worldmaking" (Goodman, 1978) are temporally structured ways of world-making is that as Dasein we are temporal beings through and through. Human subjectivity—this is Heidegger's basic idea—can be described as the execution of time. Dasein "*exists as* the primordial temporalizing of temporality" (Heidegger, 1993, p. 486). Heidegger attempts to describe this occurrence of subjectivity's self-constitution in the temporalizing of time using the temporal double movement adopted from Kierkegaard. Unlike Kierkegaard, however, for whom the double movement of human existence only fails to lead us to desperation when it occurs with conscious belief in God, Heidegger considers successful temporal self-fulfilment to be possible in the absence of divine transcendence. Although Heidegger—as Kierkegaard had already done in his speech *At a Graveside* (Kierkegaard, 1981)—also describes the anticipation of one's own future as a "Being-towards-death" (Heidegger, 1993, pp.

278ff.), he means that this anticipation of the "possibility of the measureless impossibility of existence" (Heidegger, 1993, p. 307)—which death represents—allows a kind of "authentic" existence. A kind of existence in which the experience of radical finiteness does not occasion Kierkegaardian desperation, but which instead first opens up and frees the way to shape the new horizon of manifold possibilities within which our everyday Dasein has always been organized, without our having become aware of the character of its essential possibilities. This radical outlook towards "the future as coming towards" (*Zu-kunft*) in the sense of anticipating one's own death as the "ownmost, non-relational possibility, which is not to be outstripped, and certain" (Heidegger, 1993, p. 310) is hence also understood by Heidegger as being the self's own "resoluteness" towards itself: as authentic "potentiality-for-Being-one's-Self" (*Selbstseinkönnen*; Heidegger, 1993, p. 312).

Heidegger contrasts this distinguished basic form of human temporality with the negative image of what he calls our "everyday understanding of time" (Heidegger, 1993, p. 278). He attempts to show how the everyday understanding of time arose as a derivative of the primordial temporality of human Dasein. Or, to put it another way, Heidegger's goal is to show why and how the objectivized time we read off our clocks and calendars, and which we encounter as if it were a subject-independent reality, is brought about by the temporal processes of our self-constitution, that is, by the authentic temporality of the double-movement of human existence. Heidegger's idea is that we can hold ourselves only temporarily—in distinguished moments of our Dasein—in the authentic temporality, or the resolute anticipation of death. As a rule and in the normal run of things we anticipate a future whose content we determine with our concrete needs and plans, and whose final horizon, death, we exclude. This reduced, usual practical-everyday and convenient form of double-movement is what Heidegger calls "inauthentic temporality" (Heidegger, 1993, p. 378).

Inauthentic temporality differs once again from what Heidegger calls "*vulgäres Zeitverständnis*," a vulgar or "ordinary conception of time" (Heidegger, 1993, p. 39 and passim).[10] Whilst in the inauthentic, practical-everyday temporality a reflection of "the ecstatical constitution of temporality" (Heidegger, 1993, p. 461) remains to be sensed, in the vulgar conception of time the temporal origin of time of the temporality of human Dasein is completely obscured. Heidegger makes this difference quite clear by considering our use of clocks. He refers here to a paradox that time managers and time economists have yet to overcome. This paradox lies in that "precisely that Dasein which reckons with time and lives with its watch in its hands . . . constantly says 'I have no time' " (Heidegger, 1992, p. 15). How is it that precisely the greatest strategist of time at once suffers the greatest stress due to time? Heidegger's answer is: because to the professional time manager time has congealed into a pure now-sequence of interchangeable seconds, minutes, days, weeks, months and years, into a objectivized external temporal power, lying before him as an infinitely divisible, endless line which he can never really succeed in filling. Objectivized time slips through his fingers. Any time he saves through skilful time management immediately imposes itself on him again as empty and in need of being filled with work. It is no longer concrete concerns and needs which determine his time plan, rather it is empty time itself that awakens new needs and enforces its own capitalization.[11]

Whereas this form of dealing with time has increasingly become the norm in the second half of the twentieth century (cf. Rinderspacher, 1985), Heidegger was still able to view the vulgar conception of time as being an extreme case, from which inauthentic temporality could still be clearly delineated. In the practical contexts of everyday concerns time appears to be not an external, still merely physically determined power of the clock or "nature-time" (Heidegger, 1982, p. 262), but as a "world-time" (Heidegger, 1982, p. 262) built in to and determined by our everyday concerns.

Heidegger identifies the aspects of datability, tension and publicness as being the three central characteristics distinguishing inauthentic temporality from the vulgar conception of time. What matters to Heidegger can be shown particularly clearly by taking datability as an example. Whereas in the vulgar conception of time the respective "now-point" (*Jetztpunkt*; Heidegger, 1993, p. 482) is defined solely through its immanent relation to other now-points, that is, through the abstract relation of earlier/later, the now of every-day concerns is always integrated within concrete reference to daily business whose datability it serves: it is a "now that" (*Jetzt, da*; Heidegger, 1993, p. 461). In this context Heidegger highlights: "When we look at the clock and say 'now' we are not directed toward the now as such but toward that *wherefore* and *whereto* there is still time now; we are directed toward what occupies us, what presses hard upon us, what it is time for, what we want to have time for" (Heidegger, 1982, p. 259). From this he concludes: "The fact that the structure of datability belongs essentially to what has been interpreted with the "now," "then" and "on that former occasion," becomes the most elemental proof that what has thus been interpreted has originated in the temporality which interprets itself. When we say "now," we always understand a "now *that* so and so" though we do not say all of this. Why? Because the "now" interprets a *making-present* of entities. In the "now that" lies the ecstatical character of the Present. The *datability* of the "now," the "then" and the "on that former occasion," reflects the *ecstatical* constitution of temporality, and is *therefore* essential for the time itself which has been expressed" (Heidegger, 1993, p. 460f.).

In summary it can be said that in Heidegger's differentiation between authentic temporality, inauthentic temporality and the vulgar conception of time there is a continuation of the relativization of objective time—which began with Kant's distinction between time as "formal intuition" and time as a "form of intuition"—in the concrete conditions of human Being-in-the-world. This continuation has a dual aspect. On the one hand, Heidegger relativizes the objective view of time underlying the vulgar conception of time with recourse to the pragmatic, inauthentic temporality that has drifted into our everyday dealings with time in relation to daily concerns. At this level Heidegger's analysis of temporality can be read as the consistent continuation of the analysis of Dasein in part 1 of *Being and Time* which has been pragmatically interpreted by Rorty, Tugendhat and others. On the other hand, Heidegger relativizes both the objective view of time underlying the vulgar understanding of time and the pragmatic view of time underlying inauthentic temporality with recourse to the distinguished and in his view fundamental form of authentic temporality. Proceeding from this fundamental form of temporality, Heidegger thought he would be able to make the transition from the analysis of Dasein to fundamental ontology. Hence it simultaneously marks the inner turning point at which Heidegger's phenomenology of the temporality of human Dasein is subsumed within and enshrouded by the overarching fundamental ontological perspective of *Being and Time*.

From this state of affairs Tugendhat drew the double conclusion that, firstly, the temporalizing structure of authentic temporality is reducible to a simple state of affairs and that, secondly, this reduced structure is not capable of founding the other temporal forms which Heidegger describes (Tugendhat, 1992b, esp. pp. 578–580). According to the pragmatic intention both of Tugendhat's suggestions are to be complied with. The way in which they are executed and theoretically justified is to be modified critically in contrast to Tugendhat. To begin with, however, the two conclusions, as developed by Tugendhat, will be set out.

With regard to Tugendhat's first conclusion: The reduced meaning that Tugendhat is prepared to extract from Heidegger's description of the authentic future of Dasein consists in that it is concerned with the "simple verbal distinction between what we usually call the future and one's behaviour towards the future" (Tugendhat, 1992b, p. 579). What Tugendhat has in mind here, without highlighting it clearly, is not unlike the difference which Heidegger describes in his analysis as the difference between the pragmatic constitution of inauthentic temporality and the objective theoretical constitution of the vulgar concept of time. The vulgar concept of time grasps the future as a now-point located later on the objective time beam than the current now-point. In contrast to this, the pragmatic understanding of time consists of concrete behavior with regard to one's own future plans, self-projections and projects. Thus it does not comprehend the future in terms of the present as an abstract now-point, but conversely always defines the respective now-point already in terms of these concrete projects and projections. Tugendhat's reduction of authentic temporality accordingly consists not of the revelation of, so to speak, a minimal phenomenal content which is to be preserved, but of the deletion of the authentic in favor of inauthentic temporality.

With regard to Tugendhat's second conclusion: His second conclusion results from the reduction—which is to be interpreted as the deletion of authentic temporality—carried out in his first conclusion. If the temporalizing form of authentic temporality is to be dissolved into inauthentic temporality's pragmatic apprehension of time, then the latter can no longer be described as being founded within the former. Instead of this, Tugendhat thinks, it becomes clear against the background of the reduction made in his first conclusion that "referring to one's own future has already presupposed future in the normal sense" (Tugendhat, 1992b, p. 580). Accordingly, the foundational relationship between pragmatic temporality and the vulgar concept of time ought to be reversed: "One thus sees that the coming-to-oneself, future in this allegedly primordial sense, in reality already presupposes the time which Heidegger subsequently calls the 'vulgar' one, that is, the succession of events" (Tugendhat, 1992b, p. 579). In this sense Tugendhat highlights at the beginning of his essay, with recourse to McTaggart's distinction between the shifting A series determinations (past, present, future) and the static B series relations (earlier, simultaneous, later): "It seems obvious to trace the A series back to the B series by defining: present is respectively that which is simultaneous with the now-point at which the speaker finds himself, to the future belongs that part of the B series which is later than now, and that part of the B series is past which is earlier than now" (Tugendhat, 1992b, p. 576).

This definition is already found (slightly differently formulated) in McTaggart's essay "The Unreality of Time." McTaggart writes: "We perceive events in time as being present, and those are the only events which we perceive directly. And all other events in time which, by memory or inference, we believe to be real, are regarded as past or future—those earlier than the present being past, and those later than the present being future"

(McTaggart, 1908, p. 458). In contrast to Tugendhat, however, McTaggart attempts to show in his essay "that there can be no B series where there is no A series" (McTaggart, 1908, p. 461). Within the framework of his proof McTaggart highlights that with regard to the distinctions of the A series "We can, to some extent, describe them, but they cannot be defined" (McTaggart, 1908, p. 463) without reverting once again to A determinations to do this.

If one analyzes Tugendhat's attempted definition against the background of the fact brought out by McTaggart "that the A series has to be presupposed in order to account for the A series" (McTaggart, 1908, p. 468), then it becomes clear that this attempt itself leads *ad absurdum*. Tugendhat's definition of "present," on which his further definitions of "future" and "past" are based, presupposes either that one already knows what is meant by "present," in which case it is not a definition, or that one reduces the present to being a pure now-point within the objective time series, which, however, itself *per definitionem* eludes our subjective experience.[12] Hence, with McTaggart and against Tugendhat, the linking of the B series back to human Dasein's pragmatic horizon of temporality carried out by Heidegger is to be adhered to. However, the pejorative elements, which are already expressed by Heidegger's designation of the linear time series as "vulgar time," should be eliminated from this link.

This can be achieved without having to abandon (as Tugendhat does) the model of the temporal double movement. It suffices to relativize this model, that is, to understand it too temporally, historize it and strip it of its alleged authenticity and the unreasonable nature of its fundamental ontological claim. This step, not explicitly carried out by Heidegger himself, yet suggested by his analysis, marks the basic trait of a reflexive temporalization of time carried through to its final consequences.

This step, a radical temporalization of time, is implicitly anticipated above all in Heidegger's discussion of "the ideas of Count Yorck" (Heidegger, 1993, p. 349), found in §77 of *Being and Time*. Here Heidegger highlights positively: "And Yorck . . . did not hesitate to draw the final conclusion from his insight into the historicality of Dasein" (Heidegger, 1993, p. 453). As evidence Heidegger approvingly quotes from correspondence between Yorck and Dilthey: "Behaviour and historicality are like breathing and atmospheric pressure; and—this may sound rather paradoxical—it seems to me methodologically like a residue of metaphysics to not historicize one's philosophizing" (Yorck, quoted by Heidegger, 1993, p. 453).

In his essay "Heidegger, Contingency, and Pragmatism" Rorty refers explicitly to Heidegger's Yorck quote when advocating the thesis "that the historical story which he [Heidegger] told in the 1930s was already in his mind when he wrote *Being and Time*" (Rorty, 1991, p. 40). At the same time, however, Rorty highlights: "My own guess is that in the 1920s Heidegger thought that it [the average vague understanding of Being—M.S.] is ahistorical and that in the 1930s he came to think of it as historically situated" (Rorty, 1991, p. 40). A no less ambivalent image results from Rorty's detailed interpretation of selected passages from *Being and Time* and *The Basic Problems of Phenomenology*. For both the book and the lectures, held by Heidegger in Marburg in the year of *Being and Time*'s appearance, it can be said that on the one hand passages are found in them which suggest "that the 'analytic of Dasein' in *Being and Time* is most charitably and easily interpreted as an analytic of *Western* Dasein, rather than as an account of the ahistorical conditions of the occurrence of history" (Rorty, 1991, p. 41). On the other hand, however, both the book and the lectures exhibit passages which, on the contrary, make clear that Heidegger thought "that Dasein—not just Western Dasein—had a nature which *Daseinsanalytik* could expose" (Rorty, 1991, p. 41, footnote 27). Independently of this hermeneutic problem, Rorty's Heidegger interpretation in this essay (in contrast to earlier texts) tends, however, also to interpret the second division of *Being and Time* pragmatically (Rorty, 1991, esp. p. 33f.; cf. already Okrent, 1988, pp. 191–204).

If one relativizes and historizes Heidegger's temporality concept in the manner suggested, then it becomes clear that Heidegger's levelling of the B series as vulgar time in no way results from its being linked back to the pragmatic understanding of temporality. It results far more from the contrasting of both the objectivized now series (McTaggart's B series) and pragmatic temporality (McTaggart's A series) with the authentic temporality, which functions as an overarching standard of judgement.[13] A less emphatic description of this temporalizing form of Dasein—that is, one not burdened by the fundamental ontological implications which, with Heidegger, are supposed to justify the distinction of the temporalizing structure designated as "authentic temporality"—would allow the linking of pragmatic temporality back to the formal structure of the temporal double movement to be preserved without having to adopt the hierarchical structure that Heidegger sets between the temporal forms he uncovers. This modification would amount to a pluralization of Heidegger's analysis of temporality in two respects and hence complete the reflexive temporalization of time which Kant prepared and Heidegger radicalized.

The pragmatic radicalization of Heidegger's analysis of temporality would lead, firstly, to an inner pluralization insofar as the temporal forms uncovered by Heidegger would be no longer understood in a hierarchical foundational context furnished with normative implications (authentic/inauthentic). The linking of the B series to pragmatic temporality would be understood in the same sense as Heidegger himself interpreted the binding of theory and science within contexts of everyday concern in the first division of *Being and Time* (Heidegger, 1993, esp. §16, pp. 102–107, §69, pp. 401–418; cf. Gethmann, 1993, pp. 169–206). And the boundness of pragmatic temporality, fixed in content on determinate future projections, within the temporal double movement of Dasein would be understood as boundness within a horizon in terms of which future projections first become experienceable as being concrete and determinate, and in terms of which their binding nature, one lying precisely in their contingency and transitoriness, becomes understandable.

The suggested pragmatic modification in the apprehension of the temporal double movement would, secondly, link up with an external pluralization, one no longer affecting only the internal relationship between the mutually cross-referenced temporal forms described by Heidegger, but which would take notice of alternative temporalizing forms that can no longer be understood in the conditions of the "priority of the future" (Heidegger, 1993, p. 378) presupposed by Heidegger. One might think here of the broad spectrum of divergent temporal forms ranging from Kant's "reflective judgement" and Freud's "free association," Proust's "mémoire involontaire," Benjamin's "Jetztzeit" and Newman's "now," through to Lyotard's "passage" or Derrida's "écriture."

The pragmatic pluralization of Heidegger's analysis of temporality reinterprets Heidegger's considerations from the perspective of current developments in order to make them useful in contemporary contexts of discussion. At the same time this means that it takes the literal content of *Being and Time* more seriously than is permitted by the fundamental ontological spirit determining the conception of Heidegger's *Being and Time*. If one takes the fundamental ontological horizon of *Being and Time* as a basis, and disregards altogether interpretations that attempt also to reinterpret this critically (as, for example, Vattimo, 1988), then the way in which Heidegger grasps the interrelationship between the different temporal forms—namely as a hierarchical foundational context—lapses into a renewed universalization of time defined as authentic temporality.

The relativization of objective time in *Being and Time* takes place as a foundational undertaking whose aim it is to derive the vulgar apprehension of time, by way of inauthentic temporality, from primordial temporality, that is, from the authentic temporality which is to comprise the "Being of Dasein" (Heidegger, 1993, pp. 225ff.) and thus itself be set as an ahistorical *a priori*. With regard to the supposed levelling off of the primordially ecstatical temporality, which, according to Heidegger, takes place in the constitution of the vulgar apprehension of time, we read in *Being and Time*:

> But this very levelling off, in accordance with its existential meaning, is grounded in the possibility of a definite kind of temporalizing, in conformity with which temporality temporalizes as inauthentic the kind of "time" we have just mentioned. If, therefore, we demonstrate that the "time" which is accessible to Dasein's common sense is *not* primordial, but arises rather from authentic temporality, then, in accordance with the principle, "*a potiori fit denominatio*," we are justified in designating as "*primordial time*" the *temporality* which we have now laid bare. (Heidegger, 1993, p. 377)

Looking back it is to be noted that the reflexive temporalization of time in the mainstream of modern philosophy of time takes shape with different intonations with Kant, Bergson, Husserl and Heidegger respectively, yet is carried through in all radicality and in final consequence by none of the named authors. Thus a basic temporalization of time already takes place with Kant, and then in radicalized form with Heidegger. Time is consciously thought of by neither Kant nor Heidegger as being a derivative of eternity to be grasped with the traditional means of the logos along the lines of "what is?" inquiry. Instead with Kant it appears as a theoretical condition of possibility of knowledge which first permits an object to be fixed in its "is," in its "being now" altogether. Heidegger goes a step further by demonstrating that time is not only the theoretical condition of possibility for our being able to fix anything as something within "now," that is our being able to recognize it as an object, but is, moreover, also the practical condition for our being able to project a world, to stand within a life context at all, within which the cognition of objects—to which not only Kant, but also Bergson and Husserl reduce our relation to the world—can then play a secondary role. Although not in line with its original intention, Heidegger's analysis of temporality in fact executes a pragmatic turn in modern philosophy of time. This turn opens up prospects of a radical pluralization and historization of our understanding of time pointing beyond Heidegger's fundamental ontology, and forming the focus of the reflexive temporalization of time.

## The Reflexive Temporalization Tendency and Its Relation to Objective Temporalization

To conclude I would like to relate the two ways of temporalizing time described—the objective temporalization of time occurring in modern physics, and the reflexive temporalization of time occurring in modern philosophy—to each other in two respects. First it will be shown what forms the content of the respective time concepts foregrounded in the two temporalization tendencies and how these are related to one another. Then it will be broadly outlined how the different ways of temporalizing time differ at the formal level, that is, how the difference between objective and reflexive temporalization is to be understood and how the two terms "objective" and "reflexive" are related to one another.

With regard to the content of the time concepts, the difference which distinguishes Prigogine's concept of irreversible time and Heidegger's temporality from one another must first be highlighted. This difference becomes particularly clear in a passage found in Heidegger's early lecture "The Concept of Time." In this passage Heidegger himself delimits his concept of temporality from the concept of irreversible time. In doing this he ascribes the concept of irreversible time to " 'one's' time" (Heidegger, 1992, p. 17). "One's time" is the time characteristic of the everyday Dasein which is absorbed "in concern with some 'what' that is present" (Heidegger, 1992, p. 16). Everyday concern is guided by the clock for the purposes of organization and execution of its provisions: "The clock that *one* has, every clock, shows the time of being-with-one-another-in-the-world" (Heidegger, 1992, p. 17).

Heidegger writes

> If the attempt is made to derive from the time of nature what time is, then the νῦν [now] is the μέτρον [measure] of past and future. Then time is already interpreted as present, past is interpreted as no-longer-present, future as indeterminate not-yet-present: past is irretrievable, future indeterminate. For this reason everydayness speaks of itself as that within which nature is constantly encountered. That occurrences are in time means not that they have time, but that, as occurring and existing there, they are encountered as running through a present. This time of the present is explicated as a sequence constantly rolling through the now; a sequence whose directional sense is said to be singular and irreversible. Everything that occurs rolls out of an infinite future into an irretrievable past. (Heidegger, 1992, p. 18)

Such is Heidegger's description of the irreversible structure of "one's time," which runs off from the open future into the determined past.

With the structure of irreversibility in mind he continues to interpret in the same context:

> Two things are characteristic of this interpretation: (1) irreversibility; (2) homogenizing into now-points. *Irreversibility* comprises whatever remains of authentic time

> for this explication to seize upon. This is what remains of futuricity as the fundamental phenomenon of time as Dasein. This way of viewing it looks away from the future towards the present, and from out of the present its view runs after time which flees into the past. The determination of time in its irreversibility is grounded in the fact that time was reversed beforehand. *Homogenization* is an assimilation of time to space, to Presence pure and simple; it is the tendency to expel all time from itself into a present. Time becomes fully mathematized, becomes the coordinate *t* alongside the spatial coordinates *x*,*y*,*z*. Time is irreversible. This irreversibility is the sole factor by which time still announces itself in words, the sole respect in which it resists any ultimate mathematization. Before and afterwards are not necessarily earlier and later, are not ways of temporality. In the arithmetic sequence, for example, the 3 is before the 4, the 8 after the 7. Yet the 3 is not earlier than the 4 on this account. Numbers are not earlier or later, because they are not in time at all. Earlier and later are a quite determinate before and afterwards. Once time has been defined as clock time then there is no hope of ever arriving at its original meaning again. (Heidegger, 1992, p. 18f.)

In order to interpret this passage appropriately it is helpful to recall the distinctions made by Heidegger in the analysis of temporality in *Being in Time*, which is more developed than the early essay, between the vulgar understanding of time, inauthentic and authentic temporality. In the spirit of the suggested pragmatic reading of Heidegger's analysis of temporality I will use these distinctions neutrally, that is, without Heidegger's usual pejorative accent. This will be terminologically reflected in that, in place of the normatively loaded terms used by Heidegger, I will speak of the following three temporal forms: firstly, of the linear B series underlying the measurement of time with a clock; secondly, of the pragmatic temporality of the A series resulting from the concrete horizon of our concerns and projects; and, thirdly, of the temporalizing structure of the formal double movement that permits us to distance ourselves from these projects and reflectively to relativize and historize them.

Before I can apply the pragmatic instruments in the interpretation of the Heidegger passage quoted, a few comments must first be made about Prigogine and Stengers' time-theoretical convergence thesis. The convergence thesis states that through the introduction of the concept of irreversible time the time-theoretical problem dimension, which Heidegger's analysis of temporality had unfurled as being philosophical in a distinguished sense, is drawn into physics. It is obvious that for the convergence thesis formulated by Prigogine and Stengers, not the pragmatic, but the fundamental ontological reading of Heidegger's analysis of temporality would be relevant. This becomes particularly clear when one is reminded of the way in which a series of authors have sharpened Prigogine and Stengers' convergence thesis with a view to Heidegger.

So, for instance, in their joint essay "The Time Tree," in which they rely on Prigogine's research, Friedrich Cramer and Wolfgang Kaempfer suppose that with the new scientific instruments for dealing with time "time [can] in fact be proven to be the 'Being of the being' " (*Sein des Seienden*; Cramer and Kaempfer, 1990, p. 133; cf. also Cramer, 1993). Unlike Prigogine and Stengers, Cramer and Kaempfer hence interpret the current temporalization of science not only as the experimental transliteration of the temporality structures unfurled by Heidegger, but also as the redemption of the project, outlined but not carried out by Heidegger in *Being and Time*, of a fundamental ontology: as the uncovering of time as the meaning of being. In this sense Kaempfer has suggested in his book *Time and Clocks* that the gap between the analysis of temporality in Heidegger's early work and the thinking of Being in his later work might be closed through a time-philosophical evaluation of Prigogine's research (Kaempfer, 1991, pp. 13f., 85).

Manfred Eigen established the link between Prigogine's self-organization theory and Heidegger's analysis of temporality only indirectly, and to this extent more cautiously than Cramer and Kaempfer. Eigen, who was the first to introduce the term "temporality" explicitly to the description of physical and biological self-organization processes (Eigen, 1984, esp. pp. 45ff.), concludes his essay "Evolution and Temporality" with a chapter bearing the title "Being and Temporality." It begins with an indirect reference to Heidegger which simultaneously postpones decided debate over philosophical time theory: "The heading, much as it tempts us to an excursion into philosophy, is intended only to suggest that the concept of time in modern physics is closely intertwined with Being" (Eigen, 1984, p. 55). Yet, for Eigen too, this is linked with the strong convergence theory in the sense of Prigogine and Stengers. In Eigen's formulation this states that in future with the means of self-organization theory "subjective temporal experience: 'conscious experience' " will be definable using "the objective processes underlying them: catastrophes, bifurcations or instabilities" (Eigen, 1984, p. 56).

In order to meet the strong demand—made by Prigogine and Stengers, Cramer and Kaempfer, and Eigen—of being able to explain the internal intersection between the temporality of Dasein and the temporality of Being with the means of self-organization theory, the overall perspective of fundamental ontology is needed. According to the Heidegger passage quoted, however, it is precisely from this perspective that the irreversible time structure would appear as the signature of an inauthentic temporality determined in terms of "clock time" (Heidegger, 1992, p. 18), from which "there is no hope of ever arriving" at the "original meaning" (Heidegger, 1992, p. 19) of a uniform time embracing the temporality of Dasein and the temporality of Being.

How does the relationship between Prigogine's irreversible time and Heidegger's temporality present itself from the perspective of the pragmatic Heidegger interpretation? First of all it is to be highlighted from this perspective that in the passage quoted above Heidegger has a particular form of development of pragmatic temporality in mind. This, however, is a form of development which he presents not as one such form among others in the context of "The Concept of Time," but which he absolutizes altogether in the sole and central shape of "one's time." If one examines more closely the temporal structure described by Heidegger in the quoted passage, then it becomes clear that a kind of pragmatic temporality is concerned which does not genuinely understand itself as being pragmatic temporality. The subject interpreting itself within this temporal form does not interpret itself primarily in terms of the temporal dimensions of the A series constituting the individual projection of its respective life-form in terms of the future. Rather, it understands the temporal dimensions as fixed structures inscribed in the linear B series in a preordained manner: the open future as that which is later than the present, the immutable past as that which is earlier than the present.

If one establishes from this the link with the objective temporalization of time, as has taken place within modern physics, then it becomes evident how the objective structure of irreversibility that attains physical significance with Carnot, Thomson, Clausius and Boltzmann results from the characteristic asymmetry between future and past issuing from the inscription of the A series in the B series. This, however, by no means coincides with that basic structure of pragmatic temporality which actually determines our everyday dealings with our multiple, varying, constantly shifting, and complexly interconnected projections of world and self.

This plural form of dealing with time, which is beginning to characterize the everyday world in post-industrial societies at the end of the twentieth century,[14] does not harness the temporal dimensions of the A series within the fixed framework of a supposedly objectively preordained B series, but understands the A series flexibly in terms of its embedment within the temporalizing structure of the formal double movement, in the execution of which we have learned to temporalize ourselves in our cultural, technical and historical context. It is this open and creative understanding of time which distinguishes the concept of time underlying Prigogine's theory of dissipative structures from the arrow of time model introduced into physics by classical thermodynamics. Within the framework of this concept not only is the future to be described as an open structure, but the reconstruction of a system's past is also to be relativized to the different bifurcational future paths in which the system can inscribe itself and between which it establishes links, transitions and sudden syntheses.

The parallelism between pragmatic temporality—expressed in the reflexively temporalized kinds of human life-forms that can be described by means of Heidegger's analysis of temporality—and the plural temporality of chemico-physical processes brought to light by Prigogine is the result of a scientifically successful assimilation of the physical time vocabulary to those time vocabularies with which we are used to describing our pragmatic life projections and our everyday ways of time-making. The result of this assimilation says nothing about the "Temporality of Being" (Heidegger, 1993, p. 40), but merely documents the intertwinement of philosophical, physical and life-world time vocabularies (which became possible in the conditions of computer technology), which themselves have a history and which in the framework of this history were (as can be seen retrospectively: wrongly) considered incompatible. The specific basic trait of the reflexive temporalization, which suggests itself as an alternative interpretation of Prigogine's theory of irreversible processes and distinguishes this at the formal level from the objective way of temporalizing time, is expressed in all poignancy in this contingency-theoretical view of things.

In order, finally, to bring out the formal difference existing between objective and reflexive temporalization of time, it must be made clear that the reflexive temporalization tendency is a reflexive occurrence in two senses. First of all, the temporalization of time expressed in the pragmatic interpretation of Heidegger's analysis of temporality is to be designated as reflexive in a broad sense. In this broad sense the tendency within physics, designated here as objective temporalization, also assumes a reflexive component. In both the reflexive temporalization tendency and the objective temporalization an old and static time concept is understood and relativized in terms of a new and dynamic time concept. This broad concept of "reflexive" designates a kind of objective reflexivity through which time as the object of examination is related back to itself by playing out a more temporal understanding of time against a less temporal concept of time.

The narrow and stronger concept of reflexivity, foregrounded in the philosophical temporalization tendency, is to be distinguished from this broad and weaker concept of reflexivity. With the stronger concept, self-reflexivity is involved in the sense that the execution of a reflexive temporalization of time in the objective sense—common to both temporalization tendencies—is for its part understood as being historically contingent and temporally relevant. This transition is described by Heidegger in the "Concept of Time" as follows: "In order to speak in keeping with the ontological character of our theme here, we must talk temporally about time. We wish to repeat temporally the question of what time is. Time is the 'how.' If we inquire into what time is, then one may not cling prematurely to an answer (time is such and such), for this always means a 'what' " (Heidegger, 1992, p. 22). And Heidegger concludes: "The fundamental assertion that *time is temporal* is therefore the most authentic determination . . . Time itself is meaningless; time is temporal" (Heidegger, 1992, p. 20f.).

It is the combination of these two aspects which was intended when the term "reflexive temporalization" was used in the framework of the current work. Hence, the second aspect, which clearly sets the reflexive temporalization apart from the objective temporalization, is of particular importance. It makes clear that the temporalization of time in the reflexive temporalization tendency does not take place as the essentialization of time. The reflexive temporalization does not set a true and supposedly ineluctable, and hence eternal, essence of time in the place of a deficient and inadequate understanding of time. Nor does a gradually more temporal understanding of time simply set in, replacing a less temporal understanding of time. Rather the spiralling progression of objective temporalization is breached in the course of the reflexive temporalization of time by the transition from a theoretical understanding of time, which apprehends time as an objective structure, to a pragmatic understanding of time that describes time as being a contingent form in the realization of human projects in life.

## Notes

1. On the problem of time with Nietzsche see also Stambaugh, 1987.

2. The explicit criticism levelled from this perspective by Heidegger at Bergson's time theory is found in compact form in the lecture "Logic—The Question Regarding Truth" held by Heidegger in Marburg in the 1925/26 Winter Semester (Heidegger, 1976, pp. 246–250, 266ff.; cf. Heidegger, 1990, p. 262). His criticism of Husserl's time theory is found in the lecture "The Metaphysical Foundations of Logic" from the 1928 Summer Semester (Heidegger, 1984, p. 203f.). Cf. here Meyer, 1982, pp. 37–44.

3. On the temporal structure of the Greek θεωρία see Theunissen, 1970, pp. 325–365, esp. pp. 341–344. On the different accentuations of the temporal dimensions with Bergson and Heidegger see Cassirer, 1957, pp. 162–190, esp. pp. 184–190. A corresponding comparison between Husserl and Heidegger was also provided by Theunissen, 1991, pp. 339–355.

4. Bergson altered the original French title, *Sur les donées immédiates de la conscience,*

in the English and German translations to *Time and Free Will* (1911) and *Zeit und Freiheit* (1911) respectively.

5. On this see Cassirer's account, in which he brings out the way in which Bergson asserts the past at the cost of the future: Cassirer, 1957, p. 186f.

6. For an account in English of the background of the published text of the *Lectures* see Brough, 1991, pp. xi–xviii.

7. For criticism of Heidegger's terminology see Tugendhat, 1986, p. 152.

8. For pragmatist critique of analytic philosophy's use of different means to continue the transcendental philosophical tradition see Rorty, 1979, pp. 257–312; Rorty, 1989, pp. 3–22; and Rorty, 1991, pp. 50–65.

9. On the patchwork character of *Being and Time* see Bast, 1986.

10. In Macquarrie and Robinson's translation of *Being and Time*, the German "vulgär" is rendered somewhat euphemistically as "ordinary." I prefer hereafter to translate this as "vulgar" so as to retain the clarity of the distinction between this and inauthentic temporality [trans.].

11. The mechanism underlying this "pre-projection of the time to be managed" (Gadamer, 1972, p. 224) was examined by Gadamer (Gadamer, 1972, esp. pp. 223–225). The temporal paradox expressed in the statement "the more time you save, the less you have" (Kamper, 1991, p. 255) is looked at by Kamper (Kamper, 1991, pp. 255f., 289ff.).

12. See what McTaggart describes as "another consideration" at the end of his essay in (McTaggart, 1908, p. 470) with recourse to James's theory of the "specious present."

13. On the relationship between Heidegger's analysis of temporality and McTaggart's distinction between A and B series see Okrent, 1988, p. 193; Tugendhat, 1992a, p. 130. However, neither Okrent nor Tugendhat make use of the distinction between vulgar time and pragmatic temporality developed in this interpretation. Instead they oppose vulgar time—within which, on their view, McTaggart's distinction between A and B series falls as inner differentiation—with an undifferentiated concept of temporality.

14. Cf. the empirical study by Hörning, Gerhardt and Michailow centering on the "plural orientation" (Hörning, Gerhardt, and Michailow, 1995, p. 155) of new "time management techniques" (Hörning, Gerhardt, and Michailow, 1995, p. 138) which result from the "reflexive time consciousness" (Hörning, Gerhardt, and Michailow, 1995, p. 141). See also Hörning, Ahrens and Gerhardt, 1997 which builds on this. A critical analysis of this development is found under the title "Flexible: The Restructuring of Time" in Sennett, 1998, pp. 46–63.

# Bibliography

D'Alembert, Jean le Rond (1966). *Encyclopédie ou dictionnaire raisonné des sciences, des arts et des métiers*, vol. 4. Stuttgart: Frommann (original: Paris 1754).

Andronov, Aleksandr A., and Chaikin, Semen E. (1949). *Theory of Oscillations*. Princeton, N.J.: Princeton University Press.

Apel, Karl-Otto (1973). *Transformation der Philosophie*, 2 vols. Frankfurt a.M.: Suhrkamp.

Apel, Karl-Otto (1991). "Wittgenstein und Heidegger: Kritische Wiederholung und Ergänzung eines Vergleichs." In Brian McGuinness, Jürgen Habermas et al., *Der Löwe spricht . . . und wir können ihn nicht verstehen*. Frankfurt a.M.: Suhrkamp, pp. 27–68.

Aristotle (1984a). "The Physics." In *The Complete Works of Aristotle*, vol. 1, ed. Jonathan Barnes. Oxford: Oxford University Press, pp. 315–446.

Aristotle (1984b). "On the Heavens." In *The Complete Works of Aristotle*, vol. 1, ed. Jonathan Barnes. Oxford: Oxford University Press, pp. 447–511.

Atkins, Peter William (1984). *The Second Law*. New York: W. H. Freeman.

Bachelard, Gaston (1975). *La formation de l'esprit scientifique. Contribution à une psychoanalyse de la connaissance objective*. Paris: Vrin (original edition: Paris 1938).

Baert, Patrick (ed.) (1999). *Time in Modern Intellectual Thought*. Amsterdam/New York: Elsevier.

Bast, Rainer A. (1986). "Ist Heideggers 'Sein und Zeit' ein patchwork?" *Information Philosophie*, vol. 4, October, pp. 18–30.

Baumgartner, Hans-Michael (ed.) (1993). *Das Rätsel der Zeit: Philosophische Analysen*. Freiburg/Munich: Alber.

Belousov, Boris Pawlowitsch (1987a). "Eine periodische Reaktion und ihr Mechanismus." In Kuhnert and Niedersen, 1987, pp. 71–72 (original: Moscow 1959).

Belousov, Boris Pawlowitsch (1987b). "Eine periodische Reaktion und ihr Mechanismus." In Kuhnert and Niedersen, 1987, pp. 73–82 (Manuscript of 1951, first published: Gorki 1981).

Bénard, Henri (1900). "Les tourbillons cellulaires dans une nappe liquide." *Revue Generale Sciences Pures et Appliques*, vol. 12, Paris, pp. 1261–1271.

Bergson, Henri (1910). *Time and Free Will.* New York: Macmillan (French original, entitled *Essai sur les données immediates de la conscience*, Paris 1889).

Bergson, Henri (1913). *An Introduction to Metaphysics.* London: Macmillan (French original: Paris 1903).

Bergson, Henri (1988). *Matter and Memory.* Trans. Nancy Margaret Paul and W. Scott Palmer. New York: Zone Books.

Bergson, Henri (1989). *Zeit und Freiheit.* Frankfurt a.M.: Athenäum.

Bernet, Rudolf (1985). "Editorischer Bericht." In Edmund Husserl, *Texte zur Phänomenologie des inneren Zeitbewußtseins (1893–1917)*, ed. Rudolf Bernet. Hamburg: Meiner, pp. lxix–lxxiii.

Bieri, Peter (1972). *Zeit und Zeiterfahrung: Exposition eines Problembereichs.* Frankfurt a.M.: Suhrkamp.

Boehm, Rudolf (1966). "Einleitung des Herausgebers." In Edmund Husserl, *Zur Phänomenologie des inneren Zeitbewußtseins: Husserliana*, vol. 10, pp. xiii–xliii. Den Haag: Martinus Nijhoff.

Boltzmann, Ludwig (1966a). "Further Studies on the Therman Equilibrium of Gas Molecules." In Brush, 1966, pp. 88–175 (originally in: "Sitzungsberichte der Kaiserlichen Akademie der Wissenschaften," Vienna, *Mathematisch-Naturwissenschaftliche Klasse*, vol. 66, 1872, pp. 275–370).

Boltzmann, Ludwig (1966b). "On the Relation of a General Mechanical Theorem to the Second Law of Thermodynamics." In Brush, 1966, pp. 188–193 (originally in: "Sitzungsberichte der Kaiserlichen Akademie der Wissenschaften," Vienna, *Mathematisch-Naturwissenschaftliche Klasse*, vol. 75, 1877, pp. 67–73).

Boltzmann, Ludwig (1966c). "Reply to Zermelo's Remarks in the Theory of Heat." In Brush, 1966, pp. 218–228 (originally in: *Annalen der Physik und Chemie*, vol. 57, 1896, pp. 773–784).

Boltzmann, Ludwig (1966d). "On Zermelo's Paper 'On the Mechanical Explanation of Irreversible Processes.' " In Brush, 1966, pp. 238–245 (originally in: *Annalen der Physik und Chemie*, vol. 60, 1897, pp. 392ff.).

Boltzmann, Ludwig (1980). "Über die mechanische Bedeutung des zweiten Hauptsatzes der Thermodynamik." In "Sitzungsberichte der Kaiserlichen Akademie der Wissenschaften," Vienna, *Mathematisch-Naturwissenschaftliche Klasse*, vol. 53, 1866, pp. 195–220; quoted from reprint in: Kleinert, 1980, pp. 271–296.

Brandom, Robert (1983). "Heidegger's Categories in Being and Time." *Monist*, 60 (1983), pp. 387–409.

Bray, William (1921). *Journal of the American Chemical Society*, vol. 43, pp. 1262ff.

Brough, John B. (1991). "Translator's Introduction." In Husserl, 1991, pp. xi–lvii.

Brush, Stephen G. (ed.) (1966). *Kinetic Theory*, vol. 2, *Irreversible Processes.* Oxford: Pergamon.

Brush, Stephen G. (1976). *The Kind of Motion We Call Heat: A History of the Kinetic Theory of Gases in the 19th Century*, 2 vols. Amsterdam/New York/Oxford: North Holland Personal Library.

Burger, Paul (1993). *Die Einheit der Zeit und die Vielheit der Zeiten: Zur Aktualität des Zeiträtsels*. Würzburg: Königshausen and Neumann.

Cardwell, Donald Stephen Lowell (1971). *From Watt to Clausius: The Rise of Thermodynamics in the Early Industrial Age*. London: Heinemann.

Carnot, Sadi (1986). *Reflexions on the Motive Power of Fire*. Trans. and ed. Robert Fox. New York: Lilian Barber Press (French original: Paris 1824).

Cassirer, Ernst (1957). *The Philosophy of Symbolic Forms*, vol. 3, *The Phenomenology of Knowledge*. New Haven, Conn./London: Yale University Press.

Chandrasekhar, Subrahmanyan (1961). *Hydrodynamic and Hydromagnetic Stability*. Oxford: Clarendon.

Clausius, Rudolf (1864). "Über eine veränderte Form des zweiten Hauptsatzes der mechanischen Wärmetheorie." In *ders.*, *Abhandlungen über die mechanische Wärmetheorie*, vol. 1. Brunswick: Vieweg (original in: *Poggendorff's Annalen der Physik und Chemie*, vol. 93, December, 1854, pp. 481–506).

Clausius, Rudolf (1865). "Über verschiedene für die Anwendung bequeme Formen der Hauptgleichungen der mechanischen Wärmetheorie." *Vierteljahrsschrift der naturforschenden Gesellschaft* (Zurich), vol. 10, pp. 1–59.

Clausius, Rudolf (1887). "Discussionen über die vorstehend entwickelte Form der mechanischen Wärmetheorie und ihre Begründung." In *ders.*, *Die mechanische Wärmetheorie*, vol. 1. Brunswick: Vieweg.

Clausius, Rudolf (1921). *Über die bewegende Kraft der Wärme und die Gesetze, welche sich daraus für die Wärmelehre selbst ableiten lassen*. Leipzig: Akademische Verlags-Gesellschaft (original in: *Poggendorff's Annalen der Physik und Chemie*, vol. 79, March/April, 1850, pp. 368–397, 500–524).

Comte, Auguste (1974). *The Positive Philosophy*. New York: Ams Press (French original: *Cours de philosophie positive*, 6 vols., Paris, 1830–1842).

Conen, Paul F. (1964). *Die Zeittheorie des Aristoteles*. Munich: Beck.

Coveney, Peter, and Highfield, Roger (1990). *The Arrow of Time*. London: Allen.

Cramer, Friedrich (1993). *Der Zeitbaum: Grundlegung einer allgemeinen Zeittheorie*. Frankfurt a.M.: Insel.

Cramer, Friedrich, and Kaempfer, Wolfgang (1990). "Der Zeitbaum." In *Der Komet: Almanach der Anderen Bibliothek auf das Jahr 1991*, ed. Hans Magnus Enzensberger. Frankfurt a.M.: Insel, pp. 8–190.

Davies, Paul (1974). *The Physics of Time Asymmetry*. Berkely/Los Angeles.

Davies, Paul (1988). *The Cosmic Blueprint: New Discoveries in Nature's Creative Ability to Order the Universe*. New York: Simon and Schuster.

Deleuze, Gilles (1988). *Bergsonism*. New York: Zone Books (French original: Paris 1966).

Derrida, Jacques (1992). *Given Time I, Counterfeit Money*. Chicago: University of Chicago Press (French original: Paris 1991).

Derrida, Jacques (1993). *La voix et la phénomène: Introduction au problème du signe dans la phénoménologie de Husserl*. Saint Germain: Quadrige.

Donder, Théophile de, and Rysselberghe, Pierre van (1936). *Thermodynamic Theory of Affinity: A Book of Principles*. Stanford, Calif.: Stanford University Press.

Dummett, Michael (1960). "A Defense of McTaggart's Proof of the Unreality of Time." *Philosophical Review*, vol. 69, pp. 497–504.

Düsing, Klaus (1980). "Objektive und subjektive Zeit: Untersuchungen zu Kants Zeittheorie und zu ihrer modernen kritischen Rezeption." *Kant-Studien*, vol. 71, pp. 1–34.

Düsing, Klaus (1992). "Selbstbewußtseinsmodelle: Apperzeption und Zeitbewußtsein in Heideggers Auseinandersetzung mit Kant." In *Zeiterfahrung und Personalität*, ed. Forum für Philosophie Bad Homburg. Frankfurt a.M.: Suhrkamp, pp. 89–122.

Eddington, Arthur (1943). *The Nature of the Physical World*. New York: Macmillan (originally: New York/Cambridge 1928).

Eigen, Manfred (1984). "Evolution und Zeitlichkeit." In *Die Erfahrung der Zeit*, ed. Christian Link. Stuttgart: Klett-Cotta, pp. 215–237.

Ekeland, Ivar (1984). *Le calcul, l'imprévu*. Paris: Editions de Seuil.

Elkana, Yehuda (1974). *The Discovery of the Conservation of Energy*. Cambridge, Mass.: Harvard University Press.

Epstein, Irving R., Kustin, Kenneth et al. (1989). "Oszillierende chemische Reaktionen." In *Chaos und Fraktale*, ed. James P. Crutchfield, Doyne J. Farmer et al. Heidelberg: Spektrum Akademischer Verlag, pp. 72–81.

Flasch, Kurt (1993). *Was ist Zeit? Augustinus von Hippo: Das XI. Buch der Confessiones: Historisch-philosophische Studie: Text-Übersetzung-Kommentar*. Frankfurt a.M.: Klostermann.

Gadamer, Hans-Georg (1972). "Über leere und erfüllte Zeit." In *Kleine Schriften*, vol. 3. Tübingen: Mohr, pp. 221–236 (reprinted in: Zimmerli and Sandbothe, 1993, pp. 281–198).

Galilei, Galileo (1914). *Dialogues Concerning Two New Sciences*. New York: Macmillan (original: Leiden 1638).

Gent, Werner (1962). *Die Philosophie des Raumes und der Zeit: Historische, kritische und analytische Untersuchungen*, 2 vols. Hildesheim: Olms (original: Bonn 1926–1928).

Gent, Werner (1965). *Das Problem der Zeit: Eine historische und systematische Untersuchung*. Hildesheim: Olms (original: Frankfurt a.M. 1934).

Gethmann, Carl Friedrich (1974). *Verstehen und Auslegung: Das Methodenproblem in der Philosophie Martin Heideggers*. Bonn: Bouvier.

Gethmann, Carl Friedrich (1993). *Dasein: Erkennen und Handeln: Heidegger im phänomenologischen Kontext*. Berlin/New York: de Gruyter.

Gibbs, Josiah Willard (1902). *Elementary Principles in Statistical Mechanics*. New York: Scribner's.

Gimmler, Antje, Sandbothe, Mike, and Zimmerli, Walther Ch. (1997). *Die Wiederentdeckung der Zeit: Reflexionen-Analysen-Konzepte*. Darmstadt: Primus and Wissenschaftliche Buchgesellschaft.

Giroux, Laurent (1971). *Durée Pure et Temporalité: Bergson et Heidegger*. Montréal: Bellarmin.

Gödel, Kurt (1970). "A Remark about the Relationship between Relativity Theory and Idealistic Philosophy." In *Albert Einstein: Philosopher—Scientist*, ed. Paul Arthur Schilpp. The Library of Living Philosophers, vol. 7. La Salle, Ill.: Open Court, 557–562 (first published 1949).

Goodman, Nelson (1978). *Ways of Worldmaking*. Indianapolis: Harvester.

Goodman, Nelson (1984). *Ways of Worldmaking*. Indianapolis: Hackett.

Griffin, David Ray (ed.) (1986). *Physics and the Ultimate Significance of Time: Bohm, Prigogine and Process Philosophy*. New York:.

Grünbaum, Adolf (1973). *Philosophical Problems of Space and Time*, 2nd ed. Dordrecht: Reidel (original: New York 1963).

Hegel, Georg Wilhelm Friedrich (1974). *Lectures on the History of Philosophy*, vol. 3. New York: Humanities (German original: Berlin 1832).

Heidegger, Martin (1976). "Logik—Die Frage nach der Wahrheit." In *Gesamtausgabe, II. Abteilung: Vorlesungen 1923–1944*, vol. 21. Frankfurt a.M.: Klostermann.

Heidegger, Martin (1977a). "Phänomenologische Interpretation von Kants Kritik der reinen Vernunft." In *Gesamtausgabe, II. Abteilung: Vorlesungen 1923–1944*, vol. 25. Frankfurt a.M.: Klostermann.

Heidegger, Martin (1977b). "The Age of the World Picture." In *The Question Concerning Technology and Other Essays*. New York: Harper and Row, pp. 115–154.

Heidegger, Martin (1982). *The Basic Problems of Phenomenology*. Trans. Albert Hofstadter. Bloomington: Indiana University Press (German original first published in: Heidegger, *Gesamtausgabe, II. Abteilung: Vorlesungen 1923–1944*, vol. 24. Frankfurt a.M.: Klostermann, 1975).

Heidegger, Martin (1984). *The Metaphysical Foundations of Logic*. Bloomington: Indiana University Press (German original: Frankfurt a.M. 1978).

Heidegger, Martin (1990). *Kant and the Problem of Metaphysics*. Trans. Richard Taft. Bloomington: Indiana University Press (German original: Bonn 1929).

Heidegger, Martin (1992). *The Concept of Time*. Trans. William McNeill. Oxford/Cambridge, Mass.: Blackwell (German original first published: Tübingen 1989).

Heidegger, Martin (1993). *Being and Time*. Trans. John Macquarrie and Edward Robinson. Oxford/Cambridge, Mass.: Blackwell (first published 1962; German original first published: Tübingen 1927).

Helmholtz, Hermann von (1884). *Vorträge und Reden*, vol. 1. Brunswick: Vieweg.

Helmholtz, Hermann von (1962). "On the Interaction of Natural Forces." In *Helmholtz, Popular Scientific Lectures*. New York: Dover, pp. 59–92.

Hönigswald, Richard (1965). *Die Grundlagen der Denkpsychologie: Studien und Analysen*. Stuttgart: Teubner (original: Leipzig 1925).

Hopf, Eberhard (1942). In *Berichte der Math.-Phys. Akademie der Wissenschaften*, vol. 94, Leipzig, S.1 ff.

Hörning, Karl H., Ahrens, Daniela, and Gerhard, Anette (eds.) (1997). *Zeitpraktiken. Experimentierfelder der Spätmoderne*. Frankfurt a.M.: Suhrkamp.

Hörning, Karl H., Gerhardt, Anette, and Michailow, Matthias (eds.) (1995). *Time Pioneers: Flexible Working Time and New Lifestyles*. Cambridge: Polity (German original: Frankfurt a.M. 1990).

Husserl, Edmund (1991). *On the Phenomenology of the Consciousness of Internal Time*. Dordrecht/Boston/London: Kluwer.

Janich, Peter (1996). "Die Konstitution der Zeit durch Handeln und Reden." *Kodikas/Code: Ars Semeiotica*, vol. 19, no. 1–2, pp. 133–147.

Jantsch, Erich (1980). *The Self-Organizing Universe: Scientific and Human Implications of the Emerging Paradigm of Evolution*. Oxford: Pergamon.

Jantsch, Erich (1989). *The Self-Organizing Universe: Scientific and Human Implications of the Emerging Paradigm of Evolution*. Oxford/New York: Pergamon.

Jetschke, Gottfried (1989). *Mathematik der Selbstorganisation: Quantitative Theorie deterministischer und stochastischer dynamischer Systeme*. Brunswick: Vieweg.

John, Vincenz (1968). *Geschichte der Statistik: Ein quellenmäßiges Handbuch für den akademischen Gebrauch wie für den Selbstunterricht, Erster Teil*. Wiesbaden: Sändig (original: Stuttgart 1884).

Jones, Bence (1870). *The Life and Letters of Faraday*, 2 vols. London: Longmans, Green.

Kaempfer, Wolfgang (1991). *Die Zeit und die Uhren: Mit einem Beitrag von Dietmar Kamper*. Frankfurt a.M.: Suhrkamp.

Kamper, Dietmar (1991). "Umgang mit der Zeit: Paradoxe Wiederholungen." In Kaempfer, 1991, pp. 245–351.

Kant, Immanuel (1985). *Critique of Pure Reason*. Trans. Norman Kemp Smith. London: Macmillan (first published: London 1929; German original first published: Riga 1781[A]; 1787[B]).

Kierkegaard, Sören (1981). "Vier erbauliche Reden 1844: Drei Reden bei erdachten Gelegenheiten 1845." In *Gesammelte Werke*, ed. Emanuel Hirsch and Hayo Gerdes. Gütersloh: Gütersloher Verlagshaus Mohn.

Kierkegaard, Sören (1987). *Fear and Trembling*. Ed. and trans. Howard V. Hong and Edna H. Hong. Princeton, N.J.: Princeton University Press (Danish original first published: Copenhagen 1843).

Kleinert, Andreas (1980). *Physik im neunzehnten Jahrhundert*. Darmstadt: Wissenschaftliche Buchgesellschaft.

Koyré, Alexandre (1978). *Galileo Studies*. Hassocks, UK: Harvester Press (French original: Paris: 1939).

Kroes, Peter (1982). "Order and Irreversibility." *Nature and System*, vol. 4, pp. 115–129.

Kroes, Peter (1985). *Time: Its Structure and Role in Physical Theories*. Dordrecht: Reidel.

Krohn, Wolfgang, Küppers, Günter, and Nowotny, Helga (eds.) (1990). *Self-Organization: Portrait of a Scientific Revolution*. Dordrecht/Boston/London: Kluwer Academic.

Kuhn, Thomas S. (1969). "Energy Conservation As an Example of Simultaneous Discovery." In *Critical Problems in the History of Science*, ed. Marshal Clagett. Madison/Milwaukee/London: University of Wisconsin Press, pp. 321–356.

Kuhnert, Lothar, and Niedersen, Uwe (eds.) (1987). *Selbstorganisation chemischer Strukturen: Arbeiten von F. F. Runge, R. E. Liesegang, B. P. Belousov und A. M. Zhabontinsky*. Leipzig: Geest und Portig.

Leibniz, Gottfried Wilhelm (1981). *New Essays on Human Understanding*. Cambridge/New York: Cambridge University Press (French original: Amsterdam/Leipzig 1765).

Le Poidevin, Robin, and McBeath, Murray (1993). *The Philosophy of Time*. Oxford: Oxford University Press.

Loschmidt, Josef (1876). "Über den Zustand des Wärmegleichgewichts eines Systems von Körpern mit Rücksicht auf die Schwerkraft." In "Sitzungsberichte der Akademie der

Wissenschaften," Vienna, *Mathematisch-Naturwissenschaftliche Klasse*, vol. 73, pp. 128–142.

Lübbe, Hermann (1992). *Im Zug der Zeit: Verkürzter Aufenthalt in der Gegenwart*. Berlin/Heidelberg/New York: Springer.

Lyotard, Jean-François (1991). *The Inhuman: Reflections on Time*. Cambridge: Polity (French original: Paris 1988).

Macey, Samuel L. (1991). *Time: A Bibliographic Guide*. New York: Garland.

Mach, Ernst (1986). *Principles of the Theory of Heat*. Dordrecht: Reidel (German original: Leipzig 1919).

Mainzer, Klaus (1996). *Zeit: Von der Urzeit zur Computerzeit*. Munich: Beck.

Mareschal, M., and Kestemont, E. (1987). "Experimental Evidence for Convective Rolls in Finite Two-Dimensional Molecular Models." *Journal of Statistical Physics*, vol. 48, pp. 1187ff.

Martin, Gottfried (1955). *Kant's Metaphysics and Theory of Science*. Manchester: Manchester University Press.

Maxwell, James Clerk (1860). "Illustrations of the Dynamical Theory of Gases." *Philosophical Magazine*, vol. 19, pp. 19–32, vol. 20, pp. 21–37.

Maxwell, James Clerk (1871). *Theory of Heat*. New York: Appleton.

McTaggart, John M. E. (1908). "The Unreality of Time." *Mind*, vol. 17, pp. 457–474.

Merleau-Ponty, Maurice (1962). *Phenomenology of Perception*. London: Routledge (French original: Paris, 1945).

Meyer, Rudolf W. (1982). "Bergson in Deutschland: Unter besonderer Berücksichtigung seiner Zeitauffassung." In *Studien zum Zeitproblem in der Philosophie des 20. Jahrhunderts mit Beiträgen von R. W. Meyer, E. W. Orth, R. Boehm u. W. Krewani*. Freiburg/Munich: Alber.

Minkowski, Eugène (1970). *Lived Time, Phenomenological and Psychopathological Studies*. Evanston, Ill.: Northwestern University Press.

Morin, Harald (1951). "Wille und Zeit in Schopenhauers Philosophie." *Theorie*, vol. 17, pp. 155–175.

Moscovici, Serge (1977). *Essai sur l'histoire humaine de la nature*. Paris: Flammarion (original: Paris 1968).

Most, Otto J. (1977). *Zeitliches und Ewiges in der Philosophie Nietzsches und Schopenhauers*. Frankfurt a.M.: Klostermann.

Musil, Robert (1954). *The Man without Qualities*, vol. 2. London: Secker and Warburg (German original: Berlin 1930–1933).

Newton, Isaac (1952). *Opticks*. New York: Dover.

Newton, Isaac (1968). *The Mathematical Principles of Natural Philosophy*. London: Dawsons.

Nicolis, Grégoire, and Prigogine, Ilya (1987). *Die Erforschung des Komplexen: Auf dem Weg zu einem neuen Verständnis der Naturwissenschaften*. Munich: Piper.

Nietzsche, Friedrich (1980). *Sämtliche Werke: Kritische Studienausgabe (KSA)*, vol. 11. Munich: Deutscher Taschenbuch Verlag and de Gruyter.

Oger, Erik (1991). "Einleitung." In Bergson, 1991, pp. ix–lvii.

Okrent, Mark (1988). *Heidegger's Pragmatism*. Ithaca, N.Y.: Cornell University Press.

Onsager, Lars (1931). In *Phys. Rev.*, vol. 37, pp. 405ff., and vol. 38, pp. 2265ff.

Ostwald, Wilhelm (1890). "Über Autokatalyse." In "Ber. Verh. Kgl. Sächs. Ges. Wiss.," Leipzig, *Math.-Phys. Klasse*, vol. 42, pp. 189–191.

Pflug, Günther (1959). *Henri Bergson: Quellen und Konsequenzen einer induktiven Metaphysik*. Berlin: de Gruyter.

Pieper, Hans-Joachim (1993). *Zeitbewußtsein und Zeitlichkeit: Vergleichende Analysen zu Edmund Husserls* Vorlesungen zur Phänomenologie des inneren Zeitbewußtseins *und Maurice Merleau-Pontys* Phänomenologie der Wahrnehmung. Frankfurt a.M.: Lang.

Poincaré, Henri (1966a). "On the Three-Body Problem and the Equations of Dynamics." In Brush, 1966, pp. 194–203 (originally in: *Acta math.*, vol. 13, 1890, pp. 1–270).

Poincaré, Henri (1966b). "Mechanism and Experience." In Brush, 1966, pp. 203–207 (originally in: *Revue de Métaphysique et de Morale*, vol. 1, 1893, pp. 534–537).

Popper, Karl R. (1992). "Boltzmann and the Arrow of Time." In *Popper, Unended Quest: An Intellectual Biography*. London: Routledge, pp. 156–162.

Prigogine, Ilya (1945). "Modération et transformations irréversibles des systèmes ouverts." *Académie Royale de Belgique, Bulletin de la Classe des Science*, vol. 31, pp. 600ff.

Prigogine, Ilya (1947). *Etude thermodynamique des phénomènes irreversibles*. Paris: Dunod.

Prigogine, Ilya (1973). "Time, Irreversibility and Structure." In *Physicist's Conception of Nature*, ed. Jagdish Mehra. Dordrecht/Boston: Reidel.

Prigogine, Ilya (1980). *From Being to Becoming: Time and Complexity in the Physical Sciences*. San Francisco: Freeman.

Prigogine, Ilya (1988). *Vom Sein zum Werden: Zeit und Komplexität in den Naturwissenschaften, überarbeitete und erweiterte Neuausgabe*. Munich: Piper.

Prigogine, Ilya (1989). "Die Wiederentdeckung der Zeit. Naturwissenschaft in einer Welt begrenzter Vorhersagbarkeit." In *Geist und Natur: Über den Widerspruch zwischen naturwissenschaftlicher Erkenntnis und philosophischer Welterfahrung*, ed. Hans-Peter Dürr and Walther Ch. Zimmerli. Bern/Munich/Vienna: Scherz, S.47–69.

Prigogine, Ilya (1997). *The End of Certainty: Time, Chaos, and the New Laws of Nature*. New York:.

Prigogine, Ilya (1998). *Die Gesetze des Chaos*. Frankfurt a.M.: Insel.

Prigogine, Ilya, and Defay, R. (1944). *Traité de Thermodynamique conformément aux méthodes de Gibbs et de De Donder*, 2 vols. Liège: Desoer.

Prigogine, Ilya, and Lefever, René (1968). In *Journal of Chemical Physics*, vol. 48, pp. 1695ff.

Prigogine, Ilya, and Pahaut, Serge (1985). "Die Zeit wiederentdecken." In *Zeit, die vierte Dimension in der Kunst*, ed. Michel Baudson. Weinheim: VCH, S.23–33.

Prigogine, Ilya, and Stengers, Isabelle (1981). *Dialog mit der Natur: Neue Wege naturwissenschaftlichen Denkens*. Munich: Piper.

Prigogine, Ilya, and Stengers, Isabelle (1985). *Order out of Chaos: Man's New Dialogue with Nature*. Toronto/New York/London/Sydney: Bantam.

Prigogine, Ilya, and Stengers, Isabelle (1988). *Entre le temps et l'éternité*. Paris: Fayard.

Prigogine, Ilya, and Stengers, Isabelle (1990). "Entwicklung und Irreversibilität." *Selbstor-*

*ganisation. Jahrbuch für Komplexität in den Natur-, Sozial- und Geisteswissenschaften*, vol. 1, pp. 3–18.

Prigogine, Ilya, and Stengers, Isabelle (1993). *Das Paradox der Zeit: Zeit, Chaos und Quanten*. Munich: Piper.

Prigogine, Ilya, Stengers, Isabelle, and Pahaut, Serge (1979). "La Dynamique—de Leibniz à Lucrece." *Critique*, no. 380, January, pp. 35–55 (Engl.: "Dynamics from Leibniz to Lucretius," in Michel Serres, *Hermes: Literature, Science, Philosophy*, Baltimore, Md.: Johns Hopkins University Press, 1982, pp. 137–155).

Proust, Marcel (1960). *Remembrance of Things Past*, 12 vols. London: Chatto and Windus (French original: Paris 1913–1927).

Quételet, Adolphe Lambert Jacques (1835). *Sur l'homme et le développement de ses facultés, ou essai de physique sociale*. Paris: Bachelier.

Reichenbach, Hans (1957). *The Philosophy of Space and Time*. New York: Dover (German original 1928).

Reichenbach, Hans (1971). *The Direction of Time*. Berkeley/Los Angeles/London: University of California Press (first published 1956).

Rinderspacher, Jürgen P. (1985). *Gesellschaft ohne Zeit: Individuelle Zeitverwendung und soziale Organisation der Arbeit*. Frankfurt: Campus.

Rorty, Richard (1979). *Philosophy and the Mirror of Nature*. Princeton, N.J.: Princeton University Press.

Rorty, Richard (1984). "Heidegger wider die Pragmatisten." *Neue Hefte für Philosophie*, vol. 23, pp. 1–22.

Rorty, Richard (1989). *Contingency, Irony, and Solidarity*. Cambridge: Cambridge University Press.

Rorty, Richard (1991). *Essays on Heidegger and Others, Philosophical Papers*, vol. 2. Cambridge: Cambridge University Press.

Rorty, Richard (1994). *Hoffnung statt Erkenntnis: Eine Einführung in die pragmatische Philosophie*. Vienna: Passagen.

Rorty, Richard (1995). "Philosophy and the Future." In *Rorty and Pragmatism*, ed. Herman J. Saatkamp. Nashville/London: Vanderbilt University Press.

Sambursky, Shmuel (ed.) (1978). *Der Weg der Physik: 2500 Jahre physikalischen Denkens*. Munich: Deutscher Taschenbuch Verlag (original: Zurich 1975).

Sandbothe, Mike (1989). "Schopenhauers Ästhetik: Traditionalität, Modernität, Postmodernität." In *Schopenhauer in der Postmoderne*, ed. Wolfgang Schirmacher. Schopenhauer-Studien, vol. 3. Vienna: Passagen, pp. 157–165.

Sandbothe, Mike (1994). "Die Verzeitlichung der Zeit: Grundtendenzen der modernen Zeitphilosophie und die aktuelle Wiederentdeckung der Zeit." *Glaube und Denken: Jahrbuch der Karl-Heim-Gesellschaft*, vol. 7, 108–133.

Sandbothe, Mike (1996). "Mediale Zeiten: Zur Veränderung unserer Zeiterfahrung durch die elektronischen Medien." In *Synthetische Welten: Kunst, Künstlichkeit und Kommunikationsmedien*, ed. Eckhard Hammel. Essen: Blaue Eule, pp. 133–156.

Sandbothe, Mike (1997). "Die Verzeitlichung der Zeit in der modernen Philosophie." In Gimmler, Sandbothe, and Zimmerli, 1997, pp. 41–62.

Sandbothe, Mike (1999). "Virtuelle Temporalitäten: Zeit und identitätsphilosophische Aspekte des Internet." In *Identität und Moderne*, ed. Alois Hahn and Herbert Willems. Frankfurt a.M.: Suhrkamp, S. 332–355.

Sandbothe, Mike, and Walter Ch. Zimmerli (1994). *Zeit-Medien-Wahnehmung*. Darmstadt: Wissenschaftliche Buchgesellschaft.

Sartre, Jean-Paul (1956). *Being and Nothingness: An Essay on Phenomenological Ontology*. New York: Philosophical Library (French original: Paris 1943).

Schnädelbach, Herbert (1983). *Philosophie in Deutschland 1831–1933*. Frankfurt a.M.: Suhrkamp.

Schopenhauer, Arthur (1969). *The World As Will and Representation*, vol. 1. New York: Dover (German original: Leipzig 1819).

Segrè, Emilio (1984). *From Falling Bodies to Radio Waves: Classical Physicists and Their Discoveries*. New York: W. H. Freeman.

Sennett, Richard (1998). *The Corrosion of Character: The Personal Consequences of Work in the New Capitalism*. New York/London: Norton.

Serres, Michel (1974). *Hermes III: La Traduction*. Paris: Éditions de Minuit.

Serres, Michel (1975). Introduction to *Auguste Comte: Philosophie Première*. Paris: Hermann, pp. 1–19.

Serres, Michel (1977). *Hermes IV: La Distribution*. Paris: Éditions de Minuit.

Sherover, Charles M. (1971). *Heidegger, Kant, and Time*. Bloomington: Indiana University Press.

Sherover, Charles M. (ed.) (1975). *The Human Experience of Time: The Development of Its Philosophic Meaning*. New York: New York University Press.

Sommerville, Mary (1975). *On the Connexion of the Physical Sciences*. New York: Arno (original: London 1834).

Stambaugh, Joan (1987). *The Problem of Time in Nietzsche*. London: Bucknell University Press.

Stegmaier, Werner (1987). "Zeit der Vorstellung: Nietzsches Vorstellung der Zeit." *Zeitschrift für philosophische Forschung*, vol. 41, pp. 202–228.

Stigler, Stephen M. (1987). "The Measurement of Uncertainty in Nineteenth-Century Social Science." In *The Probabilistic Revolution*, vol. 1, ed. Lorenz Krüger, Lorraine J. Daston, and Michael Heidelberger. Cambridge: MIT Press, pp. 287–292.

Szendrei, Eric V. (1989). "Bergson, Prigogine and the Rediscovery of Time." *Process Studies*, vol. 18, no. 3, pp. 181–193.

Theunissen, Michael (1970). *Hegels Lehre vom absoluten Geist als theologisch-politischer Traktat*. Berlin: de Gruyter.

Theunissen, Michael (1991). *Negative Theologie der Zeit*. Frankfurt a.M.: Suhrkamp.

Thomson, William (1882a). "An Account of Carnot's Theory of the Motive Power of Heat; with Numerical Results Deduced from Regnault's Experiments on Steam." In W. Thomson, *Mathematical and Physical Papers*, vol. 1. Cambridge: Cambridge University Press (originally in: *Transactions of the Royal Society of Edinburgh*, vol. 16, 1849, pp. 541–574).

Thomson, William (1882b). "On a Universal Tendency in Nature to the Dissipation of

Mechanical Energy." In W. Thomson, *Mathematical and Physical Papers*, vol. 1. Cambridge: Cambridge University Press, pp. 511–514 (originally in: *Transactions of the Royal Society of Edinburgh*, April 1852).

Thomson, William (1914). *Über die dynamische Theorie der Wärme mit numerischen Ergebnissen aus Herrn Joules Äquivalent einer thermischen Einheit und Herrn Regnaults Messungen an Dampf.* Trans. and ed. Walter Block. Leipzig/Berlin: Engelmann, 1914 (English original: "On the Dynamical Theory of Heat, with Numerical Results Deduced from Mr. Joule's Equivalent of a Thermal Unit, and M. Regnault's Observations on Steam, pts. I–III," in *Transactions of the Royal Society of Edinburgh*, vol. 20, March 1851, pp. 261–268, 289–298).

Thomson, William (1966). "The Kinetic Theory of the Dissipation of Energy." In Brush, 1966, pp. 176–187 (originally in: *Transactions of the Royal Society of Edinburgh*, vol. 8, 1874, pp. 325ff.).

Truesdell, Clifford A. (1980). *The Tragicomical History of Thermodynamics 1822–1854*. New York: Springer.

Tugendhat, Ernst (1970). *Der Wahrheitsbegriff bei Husserl und Heidegger*. Berlin: de Gruyter.

Tugendhat, Ernst (1982). *Traditional and Analytical Philosophy*. Cambridge: Cambridge University Press (German original: Frankfurt a.M. 1976).

Tugendhat, Ernst (1986). *Self-Consciousness and Self-Determination.* Cambridge: MIT Press.

Tugendhat, Ernst (1992a). "Heideggers Seinsfrage." In *Philosophische Aufsätze*. Frankfurt a.M.: Suhrkamp, pp. 108–135.

Tugendhat, Ernst (1992b). "Heidegger und Bergson über die Zeit." *Das Argument*, no. 194, September, pp. 573–584.

Vattimo, Gianni (1988). *The End of Modernity*. Cambridge: Polity (Italian original: Milan 1985).

Velarde, Manuel G., and Normand, Christiane (1980). "Konvektion." *Spektrum der Wissenschaft: Internationale Ausgabe in deutscher Sprache*, September, pp. 119–131.

Wendorff, Rudolf (1985). *Zeit und Kultur: Geschichte des Zeitbewußtseins in Europa, 3: Auflage*. Opladen: Westdeutscher Verlag (original: Opladen 1981).

Wheeler, John Archibald (1979). "Frontiers of Time." In *Problems in the Foundations of Physics: Proceedings of the International School of Physics "Enrico Fermi," Course LXXII*, ed. Guiliano Toraldo di Francia. Amsterdam/New York: North Holland, pp. 395–497.

Wieland, Wolfgang (1956). *Schellings Lehre von der Zeit: Grundlagen und Voraussetzungen der Weltalterphilosophie*. Heidelberg, Winter.

Winfree, Arthur (1987). *When Time Breaks Down*. Princeton, N.J.: Princeton University Press.

Wohlfart, Günter (1982). *Der Augenblick: Zeit und ästhetische Erfahrung bei Kant, Hegel, Nietzsche und Heidegger mit einem Exkurs zu Proust*. Freiburg/Munich: Alber.

Zermelo, Ernst (1966a). "On a Theorem of Dynamics and the Mechanical Theory of Heat." In Brush, 1966, pp. 208–217 (originally in: *Annalen der Physik und Chemie*, vol. 57, 1896, pp. 485–494).

Zermelo, Ernst (1966b). "On the Mechanical Explanation of Irreversible Processes." In Brush, 1966, pp. 229–237 (originally in: *Annalen der Physik und Chemie*, vol. 59, 1896, pp. 793ff.).

Zhabotinsky, Anatol Markovich (1987). "Eine periodische Oxydationsreaktion in flüssiger Phase." In Kuhnert and Niedersen, 1987, pp. 83–89 (original: Moscow 1964).

Zimmerli, Walther Ch., and Sandbothe, Mike (eds.) (1993). *Klassiker der modernen Zeitphilosophie*. Darmstadt: Wissenschaftliche Buchgesellschaft.

# Index of Names

# About the Author

**Mike Sandbothe** is professor for cultural studies and philosophy at the Friedrich Schiller University—Jena. He studied philosophy, literary theory, cultural and media studies in Tübingen, Berlin, and Bamberg. He has taught as a visiting scholar at Stanford University (1995) and was visiting professor at Bielefeld University (2000–2001). His forthcoming publications include *The Pragmatic Turn in Philosophy* (SUNY Press, 2002). Sandboth has a homepage at www.sandbothe.net and can be reached via Email at mike@sandbothe.net.

www.ingramcontent.com/pod-product-compliance
Ingram Content Group UK Ltd.
Pitfield, Milton Keynes, MK11 3LW, UK
UKHW041104150926
475561UK00001BA/34

9 780742 512900